Editor-in-Chief

Prof. Janusz Kacprzyk
Systems Research Institute
Polish Academy of Sciences
ul. Newelska 6
01-447 Warsaw
Poland
E-mail: kacprzyk@ibspan.waw.pl

For further volumes:
http://www.springer.com/series/7092

Sanchita Ghosh and Amit Konar

Call Admission Control in Mobile Cellular Networks

 Springer

Authors
Dr. Sanchita Ghosh
Artificial Intelligence Laboratory
Department of Electronics and
    Telecommunication Engineering
Jadavpur University
Calcutta
India

Prof. Amit Konar
Artificial Intelligence Laboratory
Department of Electronics and
    Telecommunication Engineering
Jadavpur University
Calcutta
India

ISSN 1860-949X   e-ISSN 1860-9503
ISBN 978-3-642-43453-2   ISBN 978-3-642-30997-7 (eBook)
DOI 10.1007/978-3-642-30997-7
Springer Heidelberg New York Dordrecht London

Printed on acid-free paper

Springer is part of Springer Science+Business Media (www.springer.com)

Preface

A cellular communication system comprises individual base station for each cellular (typically hexagonal) partitions/cells. It also consists of a mobile switching office and several hundred thousands mobile stations in each cell. The mobile switching office establishes communication between two base stations located in the respective cells of two users. Call admission control and Dynamic Channel Assignments are generally addressed as two distinct problems in a cellular (mobile) communication system. The channel assignment problem is concerned with allocating specific channels to individual subscriber when he/she is connected with his/her peers via the network. The allocation is performed on satisfying a set of network constraints involving the network parameters. The Call admission control problem on the other hand deals with admission of new and pending calls, and also serves handoffs. Because of limited system resources, all the generated/pending calls in a given cell at a given time instance cannot be served. The aim of the thesis/book is to handle both the problems jointly.

Call Admission Control (hereafter, CAC) in mobile/cellular network, deals with automatic classification of a call into one of three possible classes: serviced/blocked/dropped by the nearest base station of the user. The decision about a call is determined by studying several dynamic parameters of the network, pertaining to existing network resources and their utilization. Further, the channels for the calls selected for being serviced are also assigned by the call admission controller.

Call admission control is one of the most effective methods for optimal resource management. Here, when a call is initiated in one cell, it will request its home cell for a channel. If home cell cannot provide any free channel, the call may shift to the neighbouring cell. If the neighbours too cannot find a suitable channel then the call is dropped. Again, if the caller, while a call is in progress, moves out of a cell to any of its neighbour, a channel of the neighbouring cell will be allocated to the call. This is known as a handoff. The calls for which no channels can be assigned are dropped.

In Code Division Multiple Access (CDMA) systems, assigning a channel means allocating the appropriate power to a requesting mobile station. Due to the sharing of spectrum, this induces interference to other users. This kind of situation requires that the interference must be below a certain level to maintain the appropriate level of communication quality. This fact is elaborated in the next paragraph.

In CDMA systems, capacity is limited only by the total level of interference from all connected users. As a result, CDMA utilizes the effect of statistical multiplexing without the complex radio channel allocation or reallocation that is required in frequency-division multiple-access (FDMA) and time-division multiple-access (TDMA) systems. However, in systems based on statistical multiplexing, there exists a trade-off relationship between the system capacity and the level of communication quality. Although the so-called graceful degradation based on this trade-off relationship is one of the most essential features of CDMA systems, the communication quality must be guaranteed to a certain level on an average. The number of simultaneous users occupying a base station (BS) therefore must be limited such that an appropriate level of communication quality can be maintained. CAC thus plays a very important role in CDMA systems because it directly controls the number of users. CAC must be designed to guarantee both a grade of service (GoS), i.e., the blocking rate, and a quality of service (QoS), i.e., the probability loss for communication quality.

Call admission control has several issues for consideration. One important issue in this regard is handoff. The handoff schemes can be classified according to the way the new channel is set up and the method with which the call is handed off from the old base station to the new one. At call-level, there are two classes of handoff schemes, namely the hard and the soft handoff.

In hard handoff, the old radio link is broken before the new radio link is established and a mobile terminal communicates at most with one base station at a time. The mobile terminal changes the communication channel to the new base station with the possibility of a short interruption of the call in progress. If the old radio link is disconnected before the network completes the transfer, the call is forced to terminate. Thus, even if idle channels are available in the new cell, a handoff call may fail if the network response time for link transfer is too long. Second generation mobile communication systems based on GSM fall in this category.

In soft handoff, a mobile terminal may communicate with the network using multiple radio links through different base stations at the same time. The handoff process is initiated in the overlapping area between cells for a short duration before the actual handoff takes place. When the new channel is successfully assigned to the mobile terminal, the old channel is released. Thus, the handoff procedure is not sensitive to link transfer time. The second and the third generation CDMA-based mobile communication systems fall under this category. Soft handoff reduces call dropping at the expense of additional overhead of keeping two channels busy for a short time for a single call. Two key issues in designing soft handoff schemes are the handoff initiation time and the size of the active set of base stations the mobile is communicating with simultaneously. This study focuses on cellular networks implementing hard handoff schemes.

The other issue that plays major role in CAC is the assignment of channels. The objective of a channel assignment algorithm is to determine a spectrum efficient allocation of channels to the cells while satisfying both the traffic demand and the electromagnetic compatibility constraints.

Existing works in CAC concentrates on specific aspects of the call admission control problem. For example, a few of these works presumes a rectangular cellular structure with four neighbouring cells only and ignores the entirety of the network, thereby restricting the CAC problem in a more localized and narrow sense rather than considering it for a bigger network. The rest considers only the call drop due to insufficient channels. Unfortunately, most of the above works consider the mobile stations either static or moving very slowly in the small area with very low traffic load. The mobility factor, comprising speed and direction of movement, has been ignored in these works. Moreover, the overall network load is also ignored and hence the handoff policies used have a partial effect on the network. Most of the above schemes are localized to a single cell. Hence the channel reuse is also not done efficiently.

In this book, we propose a scheme that takes a more wide view of the call admission control problem. Here the cells are considered to be hexagonal so as to easily track the movement of MS in the neighbourhood cells. Instead of considering a single cell scenario a more global approach has been taken up using a small network to implement the algorithms to incorporate the intercellular communication efficiently. The decision of acceptance or rejection of a call dose not only depend on the feasibility and availability of the channel but also on the speed at which the MS moves and its direction of movement. Its geographical location with respect to the base station also has a significant importance. Moreover the traffic density on a cell is also considered to be a determining factor. The reuses of the channels also make the approach more effective. The algorithms proposed in the book are based on all the factors mentioned above which more perfectly handles the real world situation.

The book includes six chapters. Chapter 1 introduces the concept of mobile network and its historical background. Different radio resource allocation techniques are discussed. A basic concept of call admission control is introduced Different handoff schemes present in CAC are introduced. Next, it gives a brief account of important call admission control schemes. The schemes are partitioned depending on the nature of QoS parameter. In this chapter, different CAC using different evolutionary computing techniques are also described in brief.

Chapter 2 provides an overview of various intelligent computing tools and techniques. It begins with a review of swarm and evolutionary algorithms with special emphasis in Genetic Algorithm, Biogeography based optimization and Particle Swarm Optimization. Next it introduces the notion of fuzzy sets and their extension in two typical reasoning techniques, popularly known as Mamdani-based and Takagi-Sugeno reasoning. The principles of uncertainty management in fuzzy sets are also briefly outlined. The chapter ends with a discussion on the scope of Fuzzy and Evolutionary algorithms in Call admission Control problems.

The current literature on mobile communication usually considers the channel assignment and the call admission control as two independent problems. However, in practice these two problems are not fully independent. Chapter 3 attempts to solve the complete problem uniquely by two alternative approaches. The first approach is concerned with the development of a fuzzy to binary mapping of the measurement variables to decision variables. The latter approach deals with fuzzy to fuzzy matching, and then employs a fuzzy threshold to transform the fuzzy decisions into binary values for execution. The performance of both the call management techniques are studied with the standard Philadelphia benchmark and the results outperform reported results on independent call admission and channel assignment problems.

In chapter 4, CAC is represented as an optimization problem, where the number of calls to be serviced is maximized and the number of call rejection is to be minimized. This optimization is done under three given conditions. First, we consider a mobile receiver, which is moving with certain velocity at a certain distance from the BS. Then it is considered that all the channel allocations will fulfil the soft constraints of channel allocation. The third consideration is the network load. In this chapter, we formulated the above as a constrained optimization problem, and attempted to solve it by Genetic Algorithm. The proposed algorithm is different from the existing ones in various ways. The solutions (chromosomes) considered here are 2-dimensional allocation vectors, which represent the whole network unlike the single cell considered in the previous works. Hence it gives a complete view of the network while providing the solution and ensures better QoS.

Chapter 5 proposes a new approach to call admission control in a mobile cellular network using Bio-geography based optimization. Existing algorithms on call admission control either ignore both variation in traffic conditions or velocity of mobile devices, or at most consider one of them. This chapter overcomes the above problems jointly by formulating call admission control as a constrained optimization problem, where the primary objective is to minimize the call drop under dynamic condition of the mobile stations, satisfying the constraints to maximize the channel assignment and minimize the dynamic traffic load in the network. The constrained objective function has been minimized using Bio-geography based optimization. Experimental results and computer simulations envisage that the proposed algorithm outperforms most of the existing approaches on call admission control, considering either of the two issues addressed above. Conclusions arrived out of the book are listed in Chapter 6. Future research directions of CAC are also outlined here.

The first author would like to pay the gratitude to her parents for their mental support and forbearance during the tough period of her days when the book was written. She would also acknowledge the mental support she received from her daughter and husband as she spent a lot many hours to complete the book project, leaving them in despair particularly in holidays and weekends. The second author remembers the support he received from his wife and son to complete the book.

Jadavpur University   Sanchita Ghosh
April 30, 2012   Amit Konar

Contents

Chapter 1
An Overview of Call Admission Control in Mobile Cellular Networks

This chapter provides a thorough overview on call admission control techniques commonly employed in mobile cellular networks. It begins with an introduction to cellular technology, and gradually explores various methods and techniques for call admission control undertaken by different research groups. Strategies of call admission control under diversity of network environments have been introduced with special reference to priority of calls, predictive nature of the network and implicitness of the network, call queuing strategy, and channel borrowing schemes. Application of soft computing techniques, including artificial neural nets, genetic algorithm, fuzzy relational approach and particle swarm optimization, in call admission control is illustrated.

1.1 History of Mobile Communication

The origin of radio communications dates back to the 19th century. In 1864 James Clerk Maxwell enunciated the well-known Maxwell Equations for electromagnetic radiation. In 1876 Alexander Graham Bell invented the telephone. In 1887 Heinrich Hertz discovered "hertzian waves" which are now called as radio waves. In 1896 Guillermo Marconi carried out the world's first radio transmission. There had been scope of simplex radio communications particularly for radio reception for common people, and duplex communication among police and investigation departments over the last 50 years. The duplex radio system, however, worked for short range communications, and was far fetched to be considered for realization in a large and global sense.

Wireless communications have changed beyond recognition over the last 15 years. The first widely used cellular mobile phones were analogue and installed initially in cars due to the bulky hardware required at the time. Shortly the Bell telephone company (US) introduced the first cellular public network AMPS (Advanced Mobile Phone Service) in 1978, after hand-held devices weighing more than 1 kg became available. The AMPS had an evolution and merging with NMT (Nordic Mobile Telephone) of Germany, and appeared later in the UK as TACS (Total Access Communication Service).

The development and deployment of second-generation systems took place from the late 1980s to the present day. These were digital rather than analogue

S. Ghosh and A. Konar: CAC in a Mobile Cellular Network, SCI 437, pp. 1–62.
springerlink.com   © Springer-Verlag Berlin Heidelberg 2013

providing the end user with supposedly better voice quality, whilst providing the operators with considerable improvements to the capacity per unit bandwidth. The other advantage of second-generation systems is their roaming ability. The pan-European standard GSM (Global System for Mobile Communications) allowed international roaming for the first time throughout Europe. Many other countries throughout the world have now adopted GSM. The USA adopted an evolutionary approach to its AMPS system, developing D-AMPS (Digital AMPS). Also, a second standard is also used in the USA (IS-95), which provides an air interface based on CDMA (Code Division Multiple Access).

Wireless standards were also developed for cordless telephone applications, where users had a personal 'base' in their homes connected to a landline. An early standard in the UK and Canada was CT2, with its digital second-generation counterpart DECT (Digital Enhanced Cordless Telephone). In the USA, there are at least two cordless standards: PACS-UB, and IS-136. PACS-UB is primarily intended for a wireless PABX scenario, with multiple ports providing overlapping coverage areas, allowing portables to switch connections frequently between ports.

The Japanese have a similar system called PHS (Personal Handy phone System). The third area attracting considerable interest is Fixed Wireless Access (FWA), or alternatively known as Wireless in the Local Loop (WLL), which is intended to replace the cable from the 'last mile to the home'. FWA will probably prove most successful is in low-density communication scenarios, where cost of cabling is relatively high, and in the developing world where cabling infrastructure is not yet in place.

Third generation systems are currently under research worldwide, and are being designed to support full multimedia access. A worldwide standard may be achieved, the so-called FPLMTS (Future Public Land Mobile Telecommunication Systems). This is currently being worked out in Europe as UMTS (Universal Mobile Telephone Standard). The above scheme is known in USA as W-CDMA (Wideband CDMA). It runs in a 5 MHz bandwidth. Another 3G standard proposed in USA is known as CDMA2000. Since their proposal, there was battle over which standard to follow.

For this, another two standards have emerged. They are EDGE (Enhanced Data rates for GSM evolution) and GPRS (General Packet Radio Service). The above two standards are called 2.5G standard. These are basically 2G standards with some modification. This evolution is still continuing and we have to see where it goes now. Fig. 1.1 shows the evolution of standards.

The basic mechanism of the communication system which will be considered is that a set of entities (users) access a common medium which, in this case, is the radio channel. This concept is depicted schematically in Fig. 1.2. The frequency spectrum or bandwidth that is allocated to a certain system is a limited resource indicated by the rectangular frame. In general, there are many co-existing wireless systems.

In order to avoid interference to and from other systems, a certain level of protection is required. This is indicated by the shaded frame. The aim is to accommodate as many simultaneous users as possible (capacity) within the limited resource. In the example shown in Fig. 1.2, 4 users are considered each of whom requires an equivalent fraction of the total radio resource (illustrated by a circle).

In a digital context, this corresponds to services which require the same information bit-rate. In Fig. 1.2, the size of the circles and hence the required radio capacity are constant. In real systems the size may be time variant. Consider, for example, a speech service and periods when a speaker is silent, there is no requirement to transmit data and thus the size of the circle would shrink to merely a single point in the space. An ideal multiple access technique supports the time variant request of radio capacity because this means that, at any given time, only those resources are allocated which are actually required. Consequently, situations are avoided where more capacity is allocated than would actually be required.

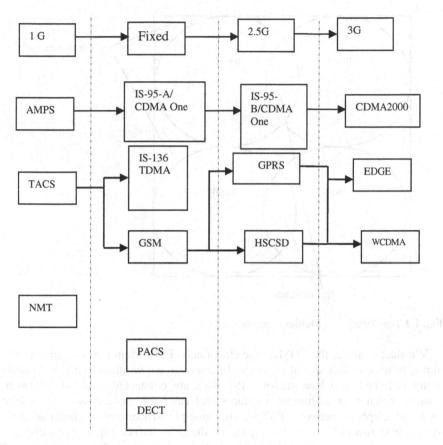

Fig. 1.1 Wireless Standards Evolution to 3rd generation

If the system is not designed carefully, users or mobile stations (MS's) will interfere with each other (gray areas). Therefore each user needs some protection which is equivalent to moving users apart. This measure, however, results in unused radio resources which, considering the immense costs for the radio frequency spectrum is inefficient. Therefore, the aim is to accommodate as many users as possible (minimizing the black colored areas) while keeping the interference at a tolerable level. The separation of users can be done in any dimensions as long as it fulfils the interference requirements. In practice the following dimensions are used:

- Frequency Division Multiple Access (FDMA),
- Time Division Multiple Access (TDMA),
- Space Division Multiple Access (SDMA),
- Code Division Multiple Access (CDMA).

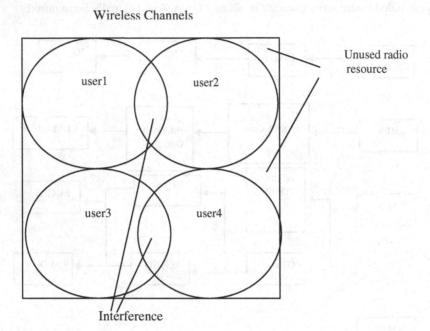

Fig. 1.2 The Principle of Multiuser access

We shall consider the CDMA case elaborately. For certain types of services the aim is to achieve full spatial coverage. In conventional wireless systems a mobile entity is linked to a base station (BS). BS's are connected to a radio network controller that uses additional interfaces that cater for the access to the public switched telephone network (PSTN). The principle structure of a cellular wireless system is shown in Fig. 1.3. The signals on the air–interface experience a distance dependent attenuation. Since the transmit powers are limited, the coverage area of a BS is limited, as well.

Due to the radial signal propagation, in theory, a single BS covers a circular area. The area that is covered by a BS is also referred to as a cell. When modeling cellular systems, cells are approximated by hexagons as they can be used to cover a plane without overlap and represent a good approximation of circles.

Since the total available radio resource is limited, the spatial dimension is used to allow wide area coverage. This is achieved by splitting the radio resource into groups. These groups are then assigned to different contiguous cells. This pattern is repeated as often as necessary until the entire area is covered. A single pattern is equivalent to a cluster.

Therefore, a radio resource which is split into i groups directly corresponds to a cell cluster of size i. In this way it is ensured that the same radio resource is only used in cells that are separated by a defined minimum distance. This mechanism is depicted in Fig. 1.4 (A group of radio resource units is indicated by a certain shade). As a consequence the separation distance grows if the cluster size increases.

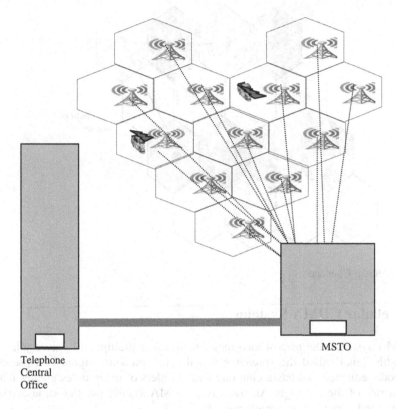

Fig. 1.3 Typical mobile communication systems

Hence, increasing the cluster size acts in favor of low interference. However, an increased cluster size means that the same radio resource is used less often within a given area. As a result, fewer users per unit area can be served. Therefore, there is a trade–off between cluster size and capacity. In an ideal scenario the total available radio resource would be used in every cell whilst the interference was kept at a tolerable level. Herein lies a particular advantage of CDMA over all other multiple access modes since the same frequency carrier can be re–used in every cell [1].

It is clear that this results in increased co–channel interference (CCI) which gradually reduces cell capacity, but the magnitude of the resulting reduction of spectral efficiency is usually less than would be obtained if a fixed frequency re–use distance was applied. The cell capacity, finally, is dependent on many system functions such as power control, handover, etc. which is why capacity in a CDMA system is described as soft–capacity. However, the fact that in a CDMA system frequency planning can be avoided may not only result in capacity gains, but it eventually makes CDMA a more flexible air interface. Next part gives some basic ideas on CDMA systems.

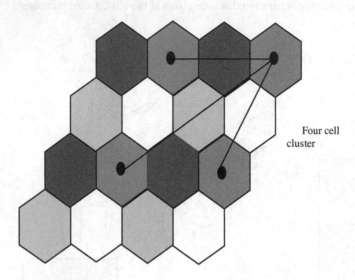

Four cell
cluster

Fig. 1.4 Cellular Concepts

1.2 Cellular CDMA Systems

In CDMA systems, the narrowband message signal is multiplied by a very large bandwidth signal called the spreading signal. The spreading signal is a pseudo noise code sequence that has a chip rate that is orders of magnitudes greater than the data rate of the message. All users in a CDMA system use the same carrier frequency and may transmit simultaneously.

Each user has its own pseudorandom codeword that is approximately orthogonal to all other code words. The receiver performs a time correlation operation to detect only the specific desired codeword. All other code words appear as noise due to de-correlation. For detection of the message signal, the receiver needs to know the codeword used by the transmitter. Each user operates independently with no knowledge of the other users.

In CDMA, the power of multiple users at a receiver determines the noise floor after de-correlation. If the power of each user within a cell is not controlled such that they do not appear equal at the base station receiver, then the near-far problem occurs.

The near-far problem occurs when many mobile users share the same channel. In general, the strongest received mobile signal will capture the demodulator at a base station. In CDMA, stronger received signal levels raise the noise floor at the base station demodulators for the weaker signals, thereby decreasing the probability that the weaker signals will be received. To combat the near-far problem, power control is used in most CDMA implementations. Power control is provided by each base station in a cellular system and assures that each mobile within the base station coverage area provides the same signal level to the base station receiver.

This solves the problem of a nearby subscriber overpowering the base station receiver and drowning out the signals of far away subscribers. Power control is implemented at the base station by rapidly sampling the radio signal strength indicator (RSSI) levels of each mobile and then sending a power change command over the forward radio link. Despite the use of power control within each cell, out-of-cell mobiles provide interference which is not under the control of the receiving base station.

Spread spectrum techniques build the foundation for CDMA. Therefore, a brief summary of spread spectrum communication is presented in the following. In a spread spectrum system, the frequency bandwidth is greater than the minimum bandwidth required to transmit the desired information. There are different methods as to how the spreading of the spectrum can be accomplished:

Direct sequence (DS) spread spectrum: A signal with a certain information bit rate is modulated on a frequency carrier with a much higher bandwidth than would be required to transmit the information signal. Each user is assigned a unique code sequence which has the property that the individual user's information can be retrieved after dispreading.

Frequency hopping (FH) spread spectrum: The available channel bandwidth is subdivided into a large number of contiguous frequency slots. The transmitted signal occupies one or more of the available frequency slots which are chosen according to a pseudo–random sequence.

Time hopping spread spectrum: A time interval which is much larger than the reciprocal of the information bit rate is subdivided into a large number of timestamps (TS). The information symbols are transmitted in a pseudo–randomly selected TS.

Chirp or pulse–FM modulation system: The frequency carrier is swept over a wide band during a given pulse interval. It is common in all spread spectrum techniques that the available bandwidth, B, is much greater than the bandwidth required transmitting a signal with an information data rate, W. The ratio B/W is the bandwidth spreading factor or processing gain, pg. The processing gain results in a interference suppression which makes spread spectrum systems highly resistant to interference or jamming. This property in particular makes spread spectrum techniques interesting for the application to wireless multiple access communication where a large number of uncoordinated users in the same geographical area access a radio frequency resource of limited bandwidth.

Using the spread spectrum technique, the number of simultaneously active users permitted is proportional to the processing gain [2]. Since the early 1980s, this has led to the development of the CDMA technology which primarily utilizes the pseudo noise (PN) DS spread spectrum technique [3]. Apart from the PN direct sequencing a second category of CDMA techniques exists. This is described as orthogonal DS–CDMA. In this thesis, PN DS–CDMA systems are considered because orthogonal CDMA systems would require an ideal channel. In addition, CDMA based standards use, at least, a PN code for the final scrambling of the transmitted data.

The wireless communication standards which utilize CDMA techniques, for example, IS–95 and UMTS use a combination of orthogonal codes and PN codes [4], but this is merely aimed to increase the robustness of the system. Since in this thesis PN DS–CDMA techniques are considered, henceforth the expression CDMA will be used to describe this particular multiple access method.

As mentioned above, the capacity calculation of a CDMA system is more complex since it is interference limited. Each user contributes to the common noise floor which is usually assumed to be Gaussian [5].Thus, interference is a most important parameter in a CDMA system and capacity analyses focus on calculating interference quantities [5]. Since interference is dependent on many factors, for example, power control, adjacent channel leakage and handover strategies to name only a few, the capacity figures can vary significantly (soft capacity).

CDMA is used in the 2nd generation mobile communication standard IS–95 which gained special interest after it had been claimed that CDMA can achieve a greater spectral efficiency than conventional FDMA and TDMA methods [5]. For example, Viterbi [6] showed that the capacity of a CDMA system can be:

$$\text{Capacity (CDMA)} \cong 1 \text{ Bit/Sector/Hz/Cell};$$

It is assumed that the voice activity of each user to be 50% and the sectorisation gain to be 4 – 6 dB. This figure was compared to the capacity of GSM (Global System for Mobile communications):

$$\text{Capacity (GSM)} \cong 1/10 \text{ Bit/Sector/Hz/Cell},$$

where, a frequency re–use factor of 1/4, was assumed. Theoretically, when considering a single cell and an AWGN (additive white Gaussian noise) channel the multiple access schemes CDMA, FDMA and TDMA are equivalent with respect to spectral efficiency [7].

Therefore, the greater spectral efficiency of CDMA systems primarily results from three basic principles:

1. The same channel is used in every cell (channel re–use factor of 10[1]),
2. Interruptions in transmission, e.g., quiet periods of a speaker, when assuming a voice service, are exploited [5].
3. Antenna sectorisation is used.

In general, the net improvement in capacity due to all the above features, of CDMA over digital FDMA or TDMA is on the order of 4 to 6 and over analog FM/FDMA it is nearly a factor of 20 [5].

However, it was demonstrated that the advantages of CDMA systems were slightly overestimated due to two basic hypotheses that usually cannot be fulfilled in a realistic environment [8]:

1. Perfect power control,
2. All MS's are allocated to the most favorable BS, i.e., the BS offering the lowest path loss.

1.3 Radio Resource Allocation Techniques

In a cellular network certain radio resources allocation methods are required to mitigate the detrimental impact of interference (Co-Channel Interface i.e. CCI and Adjacency Channel Interface i.e. ACI).

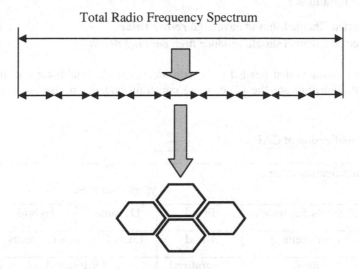

Fig. 1.5 Radio spectrum allocation

Channel Assignment Problem

The total radio spectrum allocated to a particular service producer can be divided into a set of disjoint or non-interfering radio channels (Fig. 1.5). All these channels can be used simultaneously.

The three methods used to divide spectrum into such channels are

i) **Frequency Division (FD):** Here the spectrum is divided into disjoint frequency bands.

ii) **Time Division (TD):** Here the usage of the channel is divided into disjoint time periods called time slots.

iii) **Code Division (CD):** Here the spectrum division is done using different modulation codes.

Furthermore a combination of all three can be also used to achieve desired result. Allocation of the channels among the cells should satisfy traffic demand and the electromagnetic compatibility constrains. The Constrains are categorized as soft and hard constrains. The soft constraints are describe as follows

a.　**The co-channel constraint (CCC):** where the same channel cannot be assigned to certain pairs of radio cells simultaneously.

b.　**Adjacent channel constraint (ACC):** where channels adjacent in frequency spectrum cannot de assigned to adjacent radio cells simultaneously.

c.　**Co-site Constraint (CSC):** where channel assigned in the same radio cell must have minimal separation in frequency between each other.

The hard constrains are

a.　Selected Channel should have high *re-use value*.
b.　Selected Channel should produce high *packing density*.

The major constrains that needed to be considering while establishing a channel assignment algorithm are the Co-channel constrain and the re-use value of the channels.

Table 1 Classification of CAP

Classification criteria	Types /Classes		
Co-channels Separation	Fixed	Dynamic	Hybrid
CIR measurements	Blind	Local CIR measurements	
Control	Centralized	Distributed	

Classification of Channel Allocation Scheme

Channel allocation schemes can be divided into a number of different categories depending on the comparison basis.

　　Three basic concepts of radio resource allocation are as follows:

- Static or fixed channel assignment (FCA) techniques
- Dynamic channel assignment (DCA) techniques
- Hybrid channel assignment (HCA) techniques

The principles of these methods are described in the following section.

1.3.1 Fixed Channel Assignment Techniques

An FCA method allocates a fixed fraction of all available channels to an individual cell of a cellular environment. The same group of channels is only used in cells that are separated by a minimum distance D. The channel re–use distance D ensures that CCI does not deteriorate the system performance greatly. The cluster size basically determines the system capacity, since it specifies the maximum number of simultaneously active connections that can be supported at any given time. The group size which equals the number of channels per cell, M, can be found from the relation

$$M= \text{(Available BW)} / \text{(channel BW} * \text{cluster size)}.$$

It can be seen that M is increasing with a decreasing cluster size K, but this also means that the interference is higher which, in turn, reduces the capacity or QoS. This means that a system with a greater number of channels per cell is more efficient than a system with only a few channels. This effect is well known as the trucking gain. Consequently, fixed channel assignment techniques result in poor spectral efficiency. Given that CCI varies with the cell load, there might be traffic scenarios where a lower channel re–use distance can be tolerated in favor of a temporarily higher number of channels available in a single cell (or cluster of cells).

This would require methods which dynamically monitor interference and load situations throughout the network and which carry out channel re–configurations accordingly. In contrast to DCA strategies, FCA techniques are not designed to achieve this flexibility. CDMA systems such as the UTRA–FDD interface of UMTS re–use the same channel in every cell which, in theory, makes FCA or DCA techniques superfluous, but requires special handover techniques (soft–handover). The fixed channel assignment techniques are classified as given in Fig. 1.6.

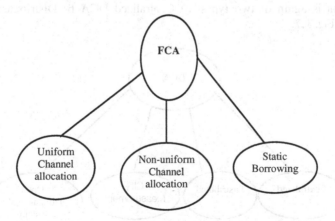

Fig. 1.6 Types of FCA

a) Uniform Channel allocation

The uniform channel allocation is efficient if the traffic distribution of the system is also uniform. Here the overall average blocking probability of the mobile system is the same as the call blocking probability in a cell. A uniform allocation of channels to cells may result in high blocking in some cells and poor channel utilization, since traffic in cellular systems can be non-uniform with temporal and spatial fluctuations.

b) Non-uniform Channel allocation

Here the number of nominal channels allocated to each cell depends on the expected traffic in that cell. So heavily loaded cells are assign more channel than the channels with comparative fewer loads.

c) Static Borrowing

In this scheme, the channels from lightly loaded cells are reassigned to the heavily loaded cells. The channels from the lightly loaded ones can be reassigned only if the distance is more than the minimum distance for reuse. This is known as static borrow since the channels can borrow free channels from its neighboring cells (donors) to accommodate new calls. When a channel is borrowed the other cells are prohibited from using it.

1.3.2 Dynamic Channel Assignment Techniques

In DCA schemes, any base station can use any channel. All channels are kept in a central pool and are assigned dynamically to new calls as they arrive in the system. After each call is completed, the channel is returned to the central pool. It is fairly straightforward to select the most appropriate channel for any call based simply on current allocation and current traffic, with the aim of minimizing the interference.

The advantage with this scheme is that channels can be moved from cells with less demand to cells with heavier demand which is time varying. Dynamic channel assignment is again of two types: a) Centralized DCA b) Distributed DCA as shown in Fig. 1.7.

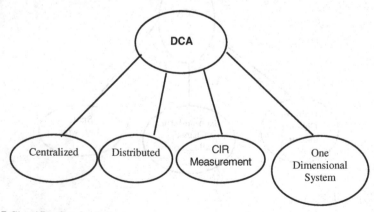

Fig. 1.7 Classification for DCA

a) **Centralized DCA:** The centralized DCA scheme involves a single controller selecting a channel for each cell. Theoretically it provides the best performance at the expense of high-centralized overhead. It is not suitable for high-density micro-cell systems. The disadvantage with this type of system is that if the main system goes down because for some reason then the whole system fails because of no other alternative.

b) **Distributed DCA:** The distributed DCA scheme involves a number of controllers scattered across the network. Here, the channel assignment decision is made by a local instance. Thus, only local information is available. Hence, the complexity is reduced considerably when this type of DCA algorithm is used.

c) **CIR measurement DCA schemes:** All mobile base station pairs are examined in channels in the same order and choose the first available with acceptable CIR

d) **One Dimension Systems:** In this scheme a mobile is assigned a channel that maximizes the minimum of the CIR's of all mobiles being served by the system at that time. A mobile is served only after all mobiles to the left of it have had a chance to be served. This sequential (left to right) order of service is chosen because it appears to be the best way for reusing the channel. The mobile immediately to the right of a given set of mobiles with channels assigned is the one that will cause the most interference at the base station servicing the given set of mobiles, and is also the one which has the most interference from that set of mobiles.

1.3.3 Hybrid Channel Assignment Techniques (HCA)

HCA schemes are the combination of both FCA and DCA techniques. In HCA schemes, the total number of channels available for service is divided into fixed and dynamic sets. The fixed set contains a number of nominal channels that are assigned to cells as in the FCA schemes and, in all cases, are to be preferred for use in their respective cells. All users share the dynamic set in the system to increase flexibility.

In a CDMA system, all users share a common channel, they are differentiated by using codes. For this, channel allocation problem is transformed to the problem of call admission control. Next part gives the idea about call admission problem.

The classification of HCA techniques are shown in Fig. 1.8.

Flexible Channel Allocation (FICA)
This scheme divides the available channels into fixed and flexible sets. Each cell is assigned a set of fixed channels that typically suffices under a light traffic load. The flexible channels are assigned to those cells whose channels have become inadequate under increasing traffic loads.

FICA techniques differ according to the time at which additional channels are assigned. The assignment of these channels among the cells is done in either a scheduled or predictive manner. In the predictive strategy, the traffic intensity or the blocking probability is constantly measured at every cell site so that the reallocation of the flexible channels can be carried at any point in time.

Fixed and Dynamic Channel Allocation

In this technique the blocking rate depending on traffic intensity. In low traffic intensity the DCA scheme is used; in heavy traffic situations the FCA strategy is used. The transition from one strategy to the other would be done gradually because a sudden transition will cause a lot of blocking.

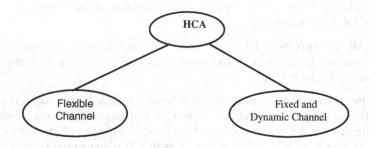

Fig. 1.8 Clasification of HCA

1.4 Call Admission Control

Dynamic channel allocation (DCA), has been extensively studied for FDMA/TDMA cellular systems as a means of increasing capacity and adapting to traffic loading variations. In DS-CDMA cellular systems, however, it is difficult to utilize DCA due to the difficulty of sharing traffic load between cells. Here, the problem of channel allocation can be viewed as call admission control.

Call admission control is one of the most effective methods for optimal resource management. When a call is initiated in one cell, it will request a channel from its home cell. In CDMA systems, assigning a channel means allocating the appropriate power to a requesting mobile. Due to the sharing of spectrum, this induces interference to other users. This kind of situation requires that the interference must be below a certain level to maintain the appropriate level of communication quality. This fact is elaborated in the next paragraph.

In CDMA systems, capacity is limited only by the total level of interference from all connected users. As a result, CDMA utilizes the effect of statistical multiplexing without the complex radio channel allocation or reallocation that is required in frequency-division multiple-access (FDMA) and time-division multiple-access (TDMA) systems. However, in systems based on statistical multiplexing, there exists a tradeoff relationship between the system capacity and the level of communication quality.

Although the so-called graceful degradation based on this tradeoff relationship is one of the most essential features of CDMA systems, the communication quality must be guaranteed to a certain level on average. The number of simultaneous users occupying a base station (BS) therefore must be limited such that an appropriate level of communication quality can be maintained. Call admission control (CAC) thus plays a very important role in CDMA systems because it directly controls the number of users. CAC must be designed to guarantee both a grade of service (GoS), i.e., the blocking rate, and a quality of service (QoS), i.e., the loss probability for communication quality.

1.4.1 Handoff Schemes

The handoff schemes can be classified according to the way the new channel is set up and the method with which the call is handed off from the old base station to the new one. At call-level, there are two classes of handoff schemes, namely the hard and the soft handoff [9-10].

1. **Hard handoff:** In hard handoff, the old radio link is broken before the new radio link is established and a mobile terminal communicates at most with one base station at a time. The mobile terminal changes the communication channel to the new base station with the possibility of a short interruption of the call in progress. If the old radio link is disconnected before the network completes the transfer, the call is forced to terminate. Thus, even if idle channels are available in the new cell, a handoff call may fail if the network response time for link transfer is too long [11]. Second generation mobile communication systems based on GSM fall in this category.

2. **Soft handoff:** In soft handoff, a mobile terminal may communicate with the network using multiple radio links through different base stations at the same time. The handoff process is initiated in the overlapping area between cells some short time before the actual handoff takes place. When the new channel is successfully assigned to the mobile terminal, the old channel is released. Thus, the handoff procedure is not sensitive to link transfer time [9], [11]. The second and third generation CDMA-based mobile communication systems fall in this category. Soft handoff decreases call dropping at the expense of additional overhead (two busy channels for a single call) and complexity (transmitting through two channels simultaneously) [11].

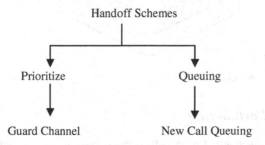

Fig. 1.9 Handoff schemes

Two key issues in designing soft handoff schemes are the handoff initiation time and the size of the active set of base stations the mobile is communicating with simultaneously [10]. This study focuses on cellular networks implementing hard handoff schemes (Fig. 1.9).

a. Prioritizing Schemes
Handoff prioritizing schemes are channel assignment strategies that allocate channels to handoff requests more readily than new calls.

b. Guard Charnels Schemes
In this scheme a number of channels are reserved exclusively for handoff calls in a cell. The remaining channels are used among the new and handoff calls.

c. Queuing Schemes
Here when the power level received by the base station in the current cell reaches a certain threshold, namely the handoff threshold, a called is placed in the queue from the neighbor cell for providing service.

The call remains in the queue until either an available channel in the new cell is found or the power by the base station in the current cell drops below a second threshold, called the receiver threshold.

d. New Call Queuing Schemes
In this method of guard channels and the queuing of new calls was introduced. This method not only minimizes blocking of handoff calls, but also increases total carried traffic.

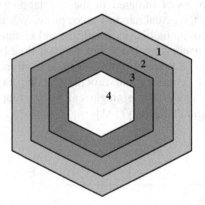

Fig. 1.10 Concentric sub cells

1.4.2 Reused Partitioning

In reused partitioning method each cell in the system is divided into two or more concentric sub cells (zones) as shown in Fig. 1.10. Because the inner zones are closer to the base station located at the center of the cell, the power level required to achieve a desired CIR in the inner zones can be much lower compared to the outer zones. The Reused Partitioning scheme can be divided as follows as given in Fig. 1.11:

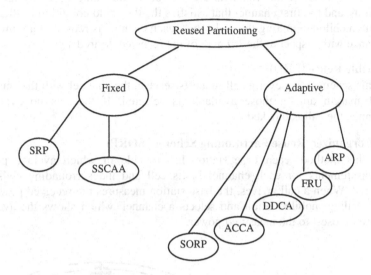

Fig. 1.11 Classification for Reused Partitioning

1.4.2.1 Fixed Reused Partitioning Scheme

In this scheme, available channels are split among several overlaid cell plans with different reuse distances.

1. **Simple Reuse Partitioning (SRP)**
   Simple RUP can be implemented by dividing the spectrum allocation into two or more groups of mutually exclusive channels. Channel assignment within the ith group is then determined by the reuse factor N_i for that group. Mobile units with the best received signal quality will be assigned to the group of channels with the smallest reuse value factor value, while those with the poorest received signal quality will be assigned to the group of channels with the largest reuse factor value.

2. **Simple Sorting Channel Assignment Algorithm (SSCAA)**
   Here each cell is divided into a number of concentric zones and assigned a number of channels. For each mobile in the cell, the base station measures the level of SIR and places the measurements in a descending order. Then it assigns channels to the set of at most mobiles with the largest values of SIR, where M is the number of available channels in the entire cell. The mobile in the set with the smallest value of SIR is assigned a channel from the outer cell zone.

1.4.2.2 Adaptive Channel Allocation Reuse Partitioning Schemes

1. **Autonomous Reuse Partitioning(ARP)**

    In this scheme all the channels are viewed in the same order by all base stations, and the first channel that satisfies the threshold condition is allocated to the mobile attempting the call. Thus, each channel is reused at a minimum distance with respect to the strength of the received desired signal.

2. **Flexible Reuse (FRU)**

    In this scheme whenever a call requests service, the channel with the smallest CIR margin among those available is selected. If there is no available channel, the call is blocked.

3. **Self-organized Reuse Partitioning Scheme (SORP)**

    In this method, each base station has a table in which average power measurements for each channel in its cell and the surrounding cells are stored. When a call arrives, the base station measures the received power of the calling mobile station and selects a channel, which shows the average power closest to the measured power.

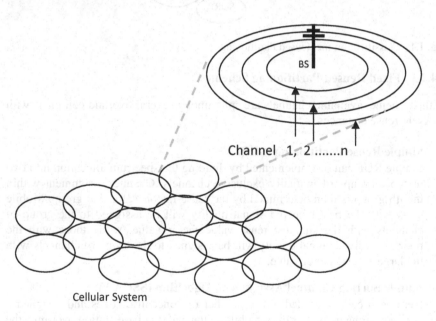

Fig. 1.12 Principle of the all-channel concentric allocation

4. **All-Channel Concentric Allocation (ACCA)**

    All radio channels of a system are allocated nominally in the same manner for each cell. Each cell is divided into N concentric regions; each region has its own channel allocation. Here, each channel is assigned a mobile belonging to the concentric region in which that channel is allocated, and has a specific desired signal level corresponding to the channel location. Therefore, each channel has its own reuse distance determined from the desired signal level (Fig. 1.12).

5. **Distributed Control Channel Allocation (DCCA)**
In this scheme all cells are identical, and channels are viewed in the same order, starting with channel number one, by all the base stations in the network. It consists of an omni-directional central station connected to six symmetrically oriented substations.

The substations are simple transceivers, and can be switched on and off under the control of the main station. When the traffic density of the cell is low, all the substations are off and the only operating station is the main station, at the center of the cell covering the entire cell area. Gradually, as call traffic increases, forced call blocking will occur due to an unacceptable level of co-channel interference o r the unavailability of resources. In this case, the main base station switches on the nearest substation to the mobile unit demanding access.

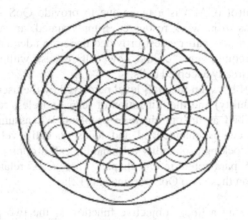

Fig. 1.13 DCCA structure

1.4.3 Performance Criteria

In this subsection, we identify some commonly used performance criteria for comparing CAC schemes. Although others exist, we will focus on the following criteria in this survey:

1. *Efficiency:* Efficiency refers to the achieved utilization level of network capacity given a specific set of QoS requirements. Scheme A is more efficient than scheme B, if the network resource utilization with scheme A is higher than that of scheme B for the same QoS parameters and the network configuration.
2. *Complexity:* Shows the computational complexity of a CAC scheme for a given network configuration, mobility patterns, and traffic parameters.

Scheme A is more complex than scheme B, if admission decision making of A involves more complex computations than scheme B.

3. *Overhead:* Refers to the signaling overhead induced by a CAC scheme on the fixed interconnection network among base stations. Some CAC schemes require some information exchange with neighboring cells through the fixed interconnection network.

4. *Adaptively:* Defined as the ability of a CAC scheme to react to changing network conditions. Those CAC schemes which are not adaptive lead to poor resource utilization.

5. *Stability:* Stability is the CAC insensitivity to short term traffic fluctuations. If an adaptive CAC reacts too fast to any load change then it may lead to unstable control. For example during a period of time all connection requests are accepted, until congestion occurs and then all requests are rejected. It is desirable that network control and management avoid such a situation.

1.5 Call Admission Control Schemes

Call admission control (CAC) is a technique to provide QoS in a network by restricting the access to network resources. Simply stated, an admission control mechanism accepts a new call request provided there are adequate free resources to meet the QoS requirements of the new call request without violating the committed QoS of already accepted calls.

There is a tradeoff between the QoS level perceived by the user (in terms of the call dropping probability) and the utilization of scarce wireless resources. In fact, CAC can be described as an optimization problem. We assume that available bandwidth in each cell is channelized and focus on call-level QoS measures. Therefore, the call blocking probability (pb) and the call dropping probability (pd) are the relevant QoS parameters in this chapter. Three CAC related problems can be identified based on these two QoS parameters [12]:

1. *MINO:* Minimizing a linear objective function of the two probabilities (pb and pd).

2. *MINB:* For a given number of channels, minimizing the new call blocking probability subject to a hard constraint on the handoff dropping probability.

3. *MINC:* Minimizing the number of channels subject to hard constraints on the new and handoff calls blocking/dropping probabilities.

As mentioned before, channels could be frequencies, time slots or codes depending on the radio technology used. Each base station is assigned a set of channels and this assignment can be static or dynamic.

1. **Deterministic CAC:** QoS parameters are guaranteed with 100% confidence [14], [15]. Typically, these schemes require extensive knowledge of the system parameters such as user mobility that is not practical, or sacrifice the scarce radio resources to satisfy the deterministic QoS bounds.

2.   **Stochastic CAC:** QoS parameters are guaranteed with some probabilistic confidence [16], by relaxing QoS guarantees, these schemes can achieve a higher utilization than deterministic approaches. Most of the CAC schemes that are investigated in this paper fall in the stochastic category. Fig. 1.14 depicts a classification of stochastic CAC schemes proposed for cellular networks. In the rest of this paper, we discuss each category in detail. In some cases, we will further expand this basic classification.

MINO tries to minimize penalties associated with blocking new and handoff calls. Thus, MINO appeals to the network provider since minimizing penalties results in maximizing the net revenue. MINB places a hard constraint on handoff call blocking thereby guaranteeing a particular level of service to already admitted users while trying to maximize the net revenue. MINC is more of a network design problem, where resources need to be allocated *a priori* based on, for example, traffic and mobility characteristics [12].

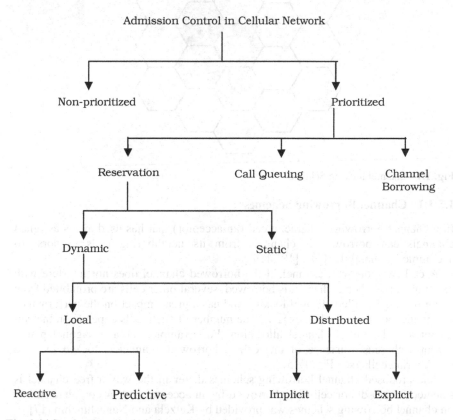

Fig. 1.14 Stochastic call admission control schemes

Since dropping a call in progress is more annoying than blocking a new call request, handoff calls are typically given higher priority than new calls in access to the wireless resources. This preferential treatment of handoffs increases the blocking of new calls and hence degrades the bandwidth utilization [13]. The most popular approach to prioritize handoff calls over new calls is by reserving a portion of available bandwidth in each cell to be used exclusively for handoffs.

In general there are two categories of CAC schemes in cellular networks:

1.5.1   Prioritization Schemes

In this section we discuss different handoff prioritization schemes, focusing on reservation schemes. Channel borrowing, call queuing and reservation are studied as the most common techniques.

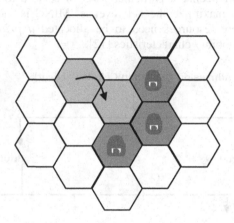

Fig. 1.15 Channel locking Scheme

1.5.1.1   Channel Borrowing Schemes

In a channel borrowing scheme, a cell (an acceptor) that has used all its assigned channels can borrow free channels from its neighboring cells (donors) to accommodate handoffs [17], [18], [19].

A cell can borrow a channel, if the borrowed channel does not interfere with existing calls. When a channel is borrowed, several other cells are prohibited from using it. This is called channel locking and has a great impact on the performance of channel borrowing schemes [20]. The number of such cells depends on the cell layout and the initial channel allocation. For example, for a hexagonal planar layout with reuse distance of one cell, a borrowed channel is locked in three neighboring cells (see Fig. 1.15).

The proposed channel borrowing schemes differ in the way a free channel is selected from a donor cell to be borrowed by an acceptor cell. A complete survey on channel borrowing schemes was provided by Katzela and Naghshinehin [17].

1.5.1.2   Call Queuing Schemes

Queuing of handoff requests in absence of channel availability can reduce the dropping probability at the expense of higher new call blocking. If the handoff attempt finds all the channels in the target cell occupied, it can be queued. If any channel is released it is assigned to the next handoff waiting in the queue.

Queuing can be done for any combination of new and handoff calls. The queue itself can be finite [21] or infinite [16]. Although finite queue systems are more realistic, systems with infinite queue are more convenient for analysis. Fig. 1.16 depicts a classification of call queuing schemes.

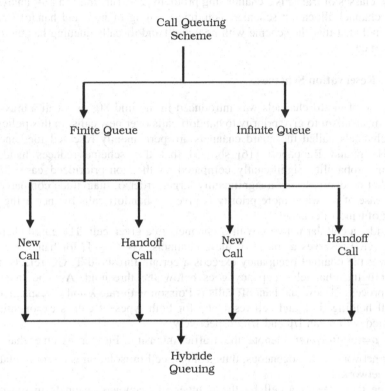

Fig. 1.16 Call queuing schemes.

Hong and Rappaport [16] analyzed the performance of the simple guard channel scheme with queuing of handoffs where handoff call attempts can be queued for the time duration in which a mobile dwells in the handoff area between cells. They used the FIFO queuing strategy and showed that queuing improves the performance of the pure guard channel scheme, i.e., probability of call drop (pd) is lower for this scheme while there is essentially no difference for probability of call block (pb).

The tolerable waiting time in queues is an important parameter. The rearranging of queued new calls due to caller impatience and the dropping of queued handoff calls as they move out of the handoff area before the handoff is accomplished successfully affect the performance of queuing schemes.

Chang et al. [21] analyzed a priority-based queuing scheme in which handoff calls waiting in queue have priority over new calls waiting in queue to gain access to available channels. They simply assumed that those calls waiting in queue cannot handoff to another cell.

Recently, Li and Chao [22] investigated a general modeling framework that can capture call queuing as well. They proved that the steady-state distribution of the equivalent queuing model has a product form solution. Queuing schemes have been mainly proposed for circuit-switched voice traffic. Their generalization to multiple classes of traffic is a challenging problem [23]. Lin and Lin [24] analyzed several channel allocation schemes including queuing of new and handoff calls. They concluded that the scheme with new and handoff calls queuing has the best performance.

1.5.1.3 Reservation Schemes

The notion of guard channels was introduced in the mid 80s as a call admission control mechanism to give priority to handoff calls over new calls. In this policy, a set of channels called the guard channels are permanently reserved for handoff calls. Hong and Rappaport [16] showed that this scheme reduces handoff-dropping probability significantly compared to the non-prioritized case. They found that pd decreases by a significantly larger order of magnitude compared to the increase of pb when more priority is given to handoff calls by increasing the number of handoff channels.

Consider a cellular network with C channels in a given cell. The guard channel scheme (GC) reserves a subset of these channels, say C − T, for handoff calls. Whenever the channel occupancy exceeds a certain threshold T, GC rejects new calls until the channel occupancy goes below the threshold. Assume that the arrival process of new and handoff calls is Poisson with rate λ and v, respectively. The call holding time and cell residency for both types of call is exponentially distributed with mean $1/\mu$ and $1/\eta$, respectively.

Let $\rho = (\lambda + v)/(\mu + \eta)$ denote the traffic intensity. Further assume that the cellular network is homogeneous, thus a single cell in isolation is a representative for the network.

Define the state of a cell by the number of occupied channels in the cell. Therefore, a continuous time Markov chain with C states can model the cell channel occupancy. The state transition diagram of a cell with C channels and C−T guard channels is shown in Fig. 1.8. Given this, it is straightforward to derive the steady-state probability P_n, that n channels are busy

$$P_n = \begin{cases} \left(\dfrac{\rho^n}{n!}\right) P_0 & 0 \le n \le T \\[3mm] \rho^T \left(\dfrac{v^{n-T}}{n!}\right) P_0 & T \le n \le C \end{cases} \qquad (1.1)$$

Where

$$P_0 = \left[\sum_{n=0}^{T} \frac{\rho^n}{n!} + \rho^T \sum_{n=T+1}^{C} \frac{v^{n-T}}{n!} \right]^{-1} \qquad (1.2)$$

And then $p_b = \sum\limits_{n=T+1}^{C} P_n$ and $p_f = P_C$ (1.3)

However, Fang and Zhang [25] showed that when the mean cell residency times for new calls and handoff calls are significantly different (as is the case for non-exponential channel holding times), the traditional one-dimensional Markov chain model may not be suitable and a two-dimensional Markov model must be applied which is more complicated. A critical parameter in this basic scheme is the optimal number of guard channels.

In fact, there is a tradeoff between minimizing pd and minimizing pb. If the number of guard channels is conservatively chosen then admission control fails to satisfy the specified pd. A static reservation typically results in poor resource utilization. To deal with this problem, several dynamic reservation schemes [17], [26–29] were proposed in which the optimal number of guard channels is adjusted dynamically based on the observed traffic load and dropping rate in a control time window.

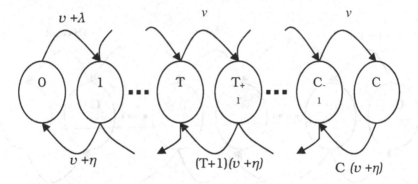

Fig. 1.17 State transition diagram of the guard channel scheme

If the observed dropping rate is above the guaranteed pd then the number of reserved channels is increased. On the other hand, if the current dropping rate is far below the target pd then the number of reserved channels is decreased. The next section investigates dynamic reservation schemes.

A different variation of the basic GC scheme is known as fractional guard channel (FGC) [12]. Whenever the channel occupancy exceeds the threshold T, the GC policy is to reject new calls until the channel occupancy goes below the threshold. In the fractional GC policy, new calls are accepted with a certain probability that depends on the current channel occupancy. Thus we have a randomization parameter which determines the probability of acceptance of a new call.

Note that both GC and FGC policies accept handoff calls as long as there are some free channels. One advantage of FGC over GC is that it distributes the newly accepted calls evenly over time which leads to a more stable control [30] where,

$$P_n = \frac{\prod\limits_{i=0}^{n-1}(v + a_i\lambda)}{(\mu + \eta)^n}P_0 \qquad 1 \leq n \leq C \qquad (1.4)$$

$$P_0 = \left[1 + \sum_{n=1}^{C}\left(\frac{\prod\limits_{i=0}^{n-1}(v + a_i\lambda)}{(\mu + \eta)^n}\right)\right]^{-1} \qquad (1.5)$$

Therefore $P_b = \sum\limits_{n=0}^{C}(1 - a_n)P_n$ and $P_f = P_C$ where $a_c = 0$.

It has been shown in [13] that due to advance reservation in reservation schemes the efficiency of cellular systems has an upper bound even if no constraint is specified on the call blocking probability. This upper bound is related to call and mobility characteristics through the mean number of handoffs per call. Moreover, the achievable efficiency decreases with decreasing cell size and with increasing call-holding time [13]

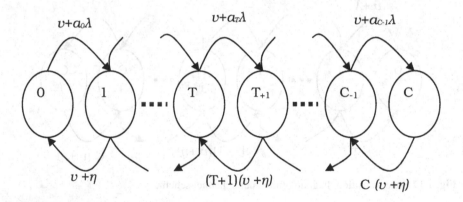

Fig. 1.18 State transition diagram of the fractional guard channel scheme

1.5.1.3.1 Dynamic Reservation Scheme

There are two approaches in dynamic reservation schemes: local and distributed (collaborative) depending on whether they use local information or gather information from neighbors to adjust the reservation threshold. In local schemes, each cell estimates the state of the network using local information only, while in distributed schemes each cell gathers network state information in collaboration with its neighboring cells.

Local Schemes

We categorize local admission control schemes into *reactive* and *predictive* schemes. By reactive approaches we refer to those admission policies that adjust their decision parameters, i.e., threshold and reservation level, as a result of an event such as call arrival, completion or rejection. Predictive approaches refer to those policies that predict future events and adjust their parameters in advance to prevent undesirable QoS degradations.

1)Reactive Approaches: The well-known guard channel (cell threshold, cut-off priority or trunk reservation) scheme (GC) is the first one in this category. GC has a reservation threshold and when the number of occupied channels reaches this threshold, no new call requests are accepted. One natural extension of this basic scheme is to use more than one threshold (e.g. two thresholds [26]) in order to have more control of the number of accepted calls. It has been shown [31] that the simple guard channel scheme performs remarkably well, often better than more complex schemes during periods in which the load does not differ from the expected level. For a discussion on different reservation strategies refer to [32] by Epstein and Schwartz.

2)Predictive Approaches: Local admission control schemes are very simple but they suffer from the lack of global information about the changes in network traffic. On the other hand, distributed admission control schemes have access to global traffic information at the expense of increased computational complexity and signaling overhead induced by information exchange between cells.

   To overcome the complexity and overhead associated with distributed schemes and benefit from the simplicity of local admission schemes, predictive admission control schemes were proposed. These schemes try to estimate the global state of the network by using some modeling/prediction technique based on information available locally. Two different approaches can be distinguished in this category:

Structural (parameter-based) modeling
The changing traffic parameters such as call arrival and departure rates are locally estimated. Assume that the control mechanism periodically measures the arrival rate. Our goal is to compute the expected arrival rate from such online measurements. Typically, a simple exponentially weighted moving average (EWMA) is used for this purpose. Let $\hat{\lambda}(i)$ and $\lambda(i)$ denote the estimated and measured new call arrival rate at the beginning of control period i, respectively. Using EWMA technique, we have

$$\hat{\lambda}(i+1) = \varepsilon\, \hat{\lambda}(i) + (1-\varepsilon)\lambda(i) \qquad (1.6)$$

where ε is the smoothing coefficient which must be properly selected. In general, a small value of ε (thus, a large value of 1- ε) can keep track of the changes more accurately, but is perhaps too heavily influenced by temporary fluctuations. On the other hand, a large value of ε is more stable but could be too slow in adapting to real traffic changes. This technique can be used to estimate the mean cell residency and call holding times as well.

Then based on these parameters, a traffic model that can describe the channel occupancy in each cell is derived. Typically, several assumptions are made about traffic parameters in this approach which are necessary to have a tractable problem.

It is clear that the EWMA in [14] is a special case of the so-called *auto regressive moving average* (ARMA) model [33] in time series analysis. There is virtually no restriction on using more complicated (and perhaps more accurate) estimation techniques.

Black-Box (measurement-based) modeling

Instead of looking at the individual components of traffic, this approach directly looks at the actual traffic. In other words, it tries to model the aggregated traffic without relying on the underlying arrival and departure processes. This approach has been proposed for multimedia systems where most of the assumptions of structural modeling are not valid [34]. The main advantage of this scheme is that it does not make any assumption about the distribution of new call arrival, handoff arrival, channel holding time and bandwidth requirements.

One of the key issues in this approach is to predict traffic in the next control time interval based on the online measurements of traffic characteristics. The goal is to forecast future traffic variations as precisely as possible, based on the measured traffic history. Traffic prediction requires accurate traffic models that can capture the statistical characteristics of actual traffic. Inaccurate models may overestimate or underestimate network traffic.

Recently, there has been a significant change in the understanding of network traffic. It has been found in numerous studies that data traffic in high-speed networks exhibits self-similarity [39–41] that cannot be captured by classical models, hence self-similar models have been developed. Among these self-similar models, fractional ARIMA [35], [36] and fractional Brownian motion [37], [38] have been widely used for network traffic modeling and prediction.

Considering that future wireless networks will offer the same services to mobile users as their wire line counterparts, it is highly possible that traffic in these networks will also exhibit self-similarity (as reported for wireless data traffic by Ziang et al. [34]). Hence, simple modeling and prediction techniques may not be accurate. Admission control based on self-similar traffic models has been already investigated for wire line networks [42], [43]. Similar approaches may be applicable to cellular communications.

1.5.1.3.2 Distributed Schemes

The fundamental idea behind all distributed schemes [27-30], [44-46] is that every mobile terminal with an active wireless connection exerts an influence upon the cells in the vicinity of its current location and along its direction of travel [27]. A group of cells, which are geographically or logically close together, form a *cluster*, as shown in Fig. 1.19. Either each mobile terminal has its own cluster independent of other terminals or all the terminals in a cell share the same cluster.

Typically, the admission decision for a connection request is made in cooperation with other cells of the cluster associated to the mobile terminal asking for admission. In Fig. 1.19(a) a cluster is defined assuming that a terminal affects all the cells in the vicinity of its current location and along its trajectory, while in Fig. 1.19(b) it is assumed that those cells that form a sector in the direction of mobile terminal's trajectory are most likely to be affected (visited) by the terminal. Fig. 1.19(c) shows a static cluster that is fixed regardless of the terminal mobility.

Each user currently in the system may either remain in the cell it is in or move to a neighboring cell; hence it can be modeled using a binomial random variable. We approximate the joint behavior of binomial distributions with a normal distribution and hence, the number of active calls in a cell at any time follows a Gaussian distribution. Also, we neglect the possibility of users having moved a distance of two or more cells and of a user arriving/completing a call during a time interval of length T.

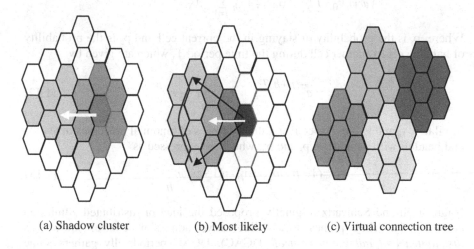

(a) Shadow cluster   (b) Most likely   (c) Virtual connection tree

Fig. 1.19 Three cluster definition.

Now, consider a hexagonal cellular system similar to those depicted in Fig. 1.19. Assume that at time $t = t_0$ a new call has arrived. New calls are admitted into the system provided that the predicted handoff failure probability of any user in the home and neighboring cells at time $t = t_0 + T$ is below the target threshold P_{QoS}. Let $n_i(t)$ denote the number of active calls in cell i at time t. Assuming that handoff failure in each cell can be approximated by the overload probability, it is obtained that

$$p_f = \Pr(n(t_0 + T) > c) \qquad (1.7)$$

Therefore the handoff failure in cell i is given by

$$p_f(i) = \frac{1}{2} erfc\left(\frac{c_i - E[n_i(t_0 + T)]}{\sqrt{2Var[n_i(t_0 + T)]}} \right) \qquad (1.8)$$

where c_i is the capacity of cell i and erfc(x) is the complementary error function defined as

$$erfc(x) = \frac{2}{\sqrt{\pi}} \int_x^{\infty} e^{-t^2} dt \qquad (1.9)$$

And the expected and variance of the number of calls at time $t_0 + T$ in cell i is given by

$$E[n_i(t_0 + T)] = n_i(t_0)p_s + p_h \sum_{j=1}^{6} n_j(t_0) \qquad (1.10)$$

$$Var[n_i(t_0 + T)] = n_i(t_0)v_s + v_h \sum_{j=1}^{6} n_j(t_0) \qquad (1.11)$$

Where, p_s is the probability of staying in the current cell and p_h is the probability of handing off to another cell during the time period T, which are given by

$$p_s = e^{-(\mu+h)T}, \quad p_h = \frac{1}{6}(1-e^{-hT}) \qquad (1.12)$$

Similarly, v_s and v_h are, respectively, the variances of binomial processes of stay and handoff with parameters p_s and p_h, which are expressed as

$$v_s = (1-p_s)p_s, \quad v_h = (1-p_h)p_h \qquad (1.13)$$

Naghshineh and Schwartz originally proposed the idea of distributed admission control [17]. They proposed a collaborative admission control known as *distributed call admission control* (DCAC). DCAC periodically gathers some information, namely the number of active calls, from the adjacent cells of the local cell to make the admission decision in combination with the local information. The analysis we presented earlier is slightly different from the original DCAC and is based on the work by Epstein and Schwartz [29]. DCAC is very restrictive in the sense that it takes into consideration information from direct neighbors only and assumes at most one handoff during the control period.

It has been shown that DCAC is not stable and violates the required dropping probability as the load increases [30]. Levin et al. [27] proposed a more complicated version of the original DCAC based on the *shadow cluster* concept, which uses dynamic clusters for each user based on its mobility pattern instead of restricting itself (as DCAC) to direct neighbors only.

A practical limitation of the shadow cluster scheme in addition to its complexity and overhead is that it requires a precise knowledge of the mobile trajectory. The so-called *active mobile probabilities* and their characterization are very crucial to the CAC algorithm. Active mobile probabilities for each user give the projected probability of being active in a particular cell at a particular instance of time.

Wu et al. [30] proposed a dynamic, distributed and stable CAC scheme called SDCA which extends the basic DCAC [17] in several ways such as using a diffusion equation to describe the evolution of the time-dependent occupancy distribution in a cell instead of the widely used Gaussian approximation. SDCA is a distributed version of the fractional guard channel in that it computes an acceptance ratio a_i for each cells i to be used for the current control period.

Consider the single-call transition probability $f_{ik}(t)$ that an ongoing call in cell i at the beginning of the control period (t = 0) is located in cell k at time t. This is in fact very similar to the active mobile probabilities introduced in [27]. For an effective control enforcing dropping probabilities in the order of 10–4 to 10–2, essentially all calls handoff successfully.

Table 2a Cluster Type vs. CAC Performance.

Cluster type	CAC efficiency	CAC complexity
Static	Moderate	Moderate
Dynamic	High	High

Wu et al. showed that for a uniform network with hexagonal cells, the probability of having n handoffs by time t, $q_n(t)$, takes the simple form

$$q_n(t) = \frac{1}{n!}\left(\frac{\eta t}{6}\right)^n e^{-(\mu+\eta)t} \tag{1.14}$$

Hence $f_{ik}(t)$ is obtained by summing over all possible paths between i and k. For example $f_{ii}(t)$ can be expressed as

$$f_{ii}(t) = q_0(t) + 6q_2(t) + 12q_3(t) + \ldots\ldots \tag{1.15}$$

Similar equations can be easily derived for $f_{ik}(t)$ [30]. Using these time-dependent transition probabilities Wu et al. computed the time-dependent mean and variance of the channel occupancy distribution, $P_{ni}(t)$, in cell i at time t. By using a diffusion approximation [47], the authors were able to find the time-dependent handoff failure, $P_{fi}(t)$, for each cell i. Hence, the average handoff failure probability over a control period of length T is found as

$$\tilde{P}_{f_i} = \frac{1}{T}\int_0^T P_{f_i}(t)\,dt \tag{1.16}$$

Finally, the acceptance ratio ai can be obtained by numerically solving the following equation [49]:

$$\tilde{P}_{f_i} = P_{Q\circ S}, \qquad 0 \le a_i \le 1 \tag{1.17}$$

A. Classification of Distributed Schemes

Distributed CAC s can be classified according to two factors:

1. Cluster definition
2. Information exchange and processing

A cluster can be either static or dynamic. In the static approach, the size and shape of the cluster is the same regardless of the network situation. In the dynamic approach however, shape and/or size of the cluster change according to the congestion level and traffic characteristics. The virtual connection tree of [46] is an example of a static cluster while the shadow cluster introduced in [27] is a dynamic cluster. A shadow cluster is defined for each individual mobile terminal based on its mobility information, e.g. trajectory, and changes as the terminal moves.

Table 2b Comparison of Dynamic CAC Schemes

	CAC scheme	Efficiency	Overhead	Complexity	Adaptively
Local	Reactive	Low	Low	Low	Moderate
	Predictive	Moderate	Low	Moderate	Moderate
Distributed	Implicit	High	Very High	High	High
	Explicit	High	High	Very High	High

It has been shown that it is not worth involving several cells in the admission control process when the network is not congested [49]. Table1a shows a tradeoff between the cluster type and the corresponding CAC performance. Typically, dynamic clusters have a better performance at the expense of increased complexity. In general, distributed CAC s can be categorized into implicit or explicit based on the involvement of cells in the decision making process:

1. *Implicit Approach:* In this approach, all the necessary information is gathered from the neighboring cells, but the processing is local. The virtual connection tree concept introduced in [46] is an example of an implicitly distributed scheme. In this scheme each connection tree consists of a specific set of base stations where each tree has a network controller. The network controller is responsible for keeping track of the users and resources. Despite the fact that information is gathered from a set of neighboring cells, the final decision is made locally in the network controller.

2. **Explicit Approach:** In this approach, not only information is gathered from the neighboring cells, but also the neighboring cells are involved in the decision making process. The shadow cluster concept introduced in [27] is an example of an explicitly distributed scheme. In this scheme a cluster of cells, the shadow cluster, is associated with each mobile terminal in a cell. Upon admitting a new call, all the cells in the corresponding cluster calculate a preliminary response that after processing by the original cell will form the final decision.

Although it is theoretically possible to involve all the network cells in the admission control process, it is expensive and sometimes useless in practice. To consider the effect of all the cells, analytical approaches involve huge matrix exponentiations. In [30] and [50] two different approximation techniques have been proposed to compute these effects with a lower computational complexity.

Table 2b shows a comparison of different dynamic CAC schemes. In general, there is a tradeoff between the efficiency and the complexity of local and distributed schemes. Table 3 compares three major distributed CAC schemes

In this table, *Naghshineh proposed basic distributed* and Schwartz [17], *shadow cluster* refers to the work of Levin et al. [27] and *stable dynamic* is due to Wu et al. [30].

Table 3 Comparison of Distributed CAC Schemes

CAC scheme	Efficiency	Complexity	Stability
Basic distributed	Moderate	Moderate	Moderate
Shadow cluster	High	High	Moderate
Stable dynamic	Very High	High	High

1.5.2 Non Prioritized Scheme

There are several schemes which functions without having any priority. The schemes are given as follows.

1.5.2.1 Optimal Control

Recall that a call admission policy is the set of decisions that indicate when a new call will be allocated a channel and when and existing call will be denied a handoff from one cell to another. Here we investigate the optimal and near-optimal admission policies proposed for three admission problems, namely,

MINO, MINB and MINC. Although optimal policies are more desirable, near-optimal policies are more useful in practice due to the complexity of optimal policies that usually leads to an intractable solution. Table IV shows a comparison of optimal and near-optimal schemes.

Decision theoretic approaches based on *Markov decision process* (MDP) [49] have been extensively studied to find the optimal CAC policy using standard optimization techniques [51].

However, for simple cases such as the one of an isolated cell in a voice system, simple Markov chains have been applied successfully [12]. A Markov decision process is just like a Markov chain, except that the transition matrix depends on the action taken by the decision maker (CAC) at each time step. The CAC receives a reward, which depends on the action and the state.

Table 4 Comparison of Optimal CAC Schemes

CAC scheme		Efficiency	Complexity
Optimal	Single service	High	High
	Multiple	High	Very High
Near-Optimal	Single service	Moderate	Low
	Multiple services	Moderate	Moderate

The goal is to find a policy, which specifies which action to take in each state, so as to maximize some function (e.g. the mean or expected sum) of the sequence of rewards. A problem formulated as an MDP can be solved iteratively [51]. This is called policy iteration, and is guaranteed to converge to the unique optimal policy. The best theoretical upper bound on the number of iterations needed by policy iteration is exponential in the number of states. However, by formulating the problem as a linear programming problem, it can be proved that one can find the optimal policy in polynomial time.

A. Optimal CAC Schemes

1. **Single Service Case:** Ramjee et al. [12] showed that the well-known GC policy is optimal for the MINO problem and a restricted version of the FGC policy is optimal for the MINB and MINC problems. In their work, a Markov chain similar to the one stated before describes channel occupancy. Although admission policies derived from the MDP formulation of the CAC [54], [55] are optimal for the MINO problem, it has been shown that a dynamic guard channel scheme is more realistic and at the same time approaches the optimal solution [55], [56].

2. **Multiple Services Case:** Introducing multiple services changes the system behavior dramatically. In contrast to single service systems, GC is no longer optimal for the MINO problem. While the optimal admission policy for single service (voice) systems is computationally complex, for multiple services (multimedia) systems it is even more complicated and expensive. In this situation, a *semi-Markov decision process* (SMDP) has been applied successfully. Optimal policies are reported for multimedia traffic in [52], [57]–[60]. In particular, Choi et al. [61] presented a centralized CAC based on SMDP, Kwon et al. [57] and Yoon et al. [62] proposed distributed CAC schemes based on SMDP, all for non-adaptive multimedia applications. Xiao et al. [58] developed an optimal scheme using SMDP for adaptive multimedia applications. Adaptive multimedia applications can change their bit-rate to adapt to network resource availability.

1.5.2.2 Near-Optimal CAC Schemes

As mentioned before, when the state of the system can be modeled as a Markov process, there exist methods to calculate the optimal call admission policy using a Markov decision process. However, for systems with a large number of states (which grows exponentially with the cell capacity and known as the curse of dimensionality) this method is impractical since it requires solving large systems of linear equations. Therefore, methods, which can calculate a near-optimal policy, are proposed in the literature. In particular, near-optimal approaches based on Markov decision processes [63], genetic algorithms [64], [65], and reinforcement learning [66] have been proposed.

1.6 Other Admission Control Scheme

There are other schemes which are also efficient in handle call admission in a network system.

1.6.1 Multiple Services Schemes

Moving from single service systems to multiple services systems raises new challenges. Particularly, wireless resource management and admission control become more crucial for efficient use of wireless resources [14], [23], [29], [67], [68]. Despite the added complexity to control mechanisms, multiple services systems are typically more flexible in terms of resource management. Usually there are some low priority services, e.g. best effort service, which can utilize unused bandwidth.

This bandwidth can be released and allocated to higher priority services upon request, e.g. when the system is fully loaded and a high priority handoff arrives. Fig. 1.20 shows a classification of guard channel based CAC schemes in single service and multiple services systems. In the figure, *multiple cutoff priority* [23] and *thinning scheme* [68] are the multiple services counterparts of GC and FGC schemes in single service systems respectively. In this context, the *thinning scheme* [68] is proposed as a generalization of the basic FGC for multiple classes' prioritized traffic. Assume that the wireless network has call requests of r priority levels and each base station has C channels. Let a_{ij} (i = 0, . . ., C and j = 1, . . . , r) denote the acceptance probabilities of prioritized classes respectively. When the number of busy channels at a base station is i, an arriving type-j call will be admitted with probability a_{ij}. All calls will be blocked when all channels are busy.

Call arrivals of priority classes are independent of each other and assumed to be Poisson with rate λj for class j. Call durations are exponentially distributed with parameter μ. A Markov chain in which the state variable is the number of busy channels in the cell can characterize this system. Let Pn denote the stationary probability at state n, $\rho_j = \dfrac{\lambda_j}{\mu}$ and $\alpha_k = \sum_{j=1}^{r} \alpha_{k,j} \rho_j$ Using balance equations we have

$$P_n = \frac{\prod_{k=0}^{n-1} \alpha_k}{n!} P_0 \tag{1.18}$$

where,

$$P_0 = \left[\sum_{n=0}^{C} \left(\frac{\prod_{k=0}^{n-1} \alpha_k}{n!} \right) \right]^{-1} \tag{1.19}$$

Then the blocking probability for class *j* is given by

$$P_b{}^j = \sum_{i=T+1}^{C} \left(1 - \alpha_{ij}\right) P_i \tag{1.20}$$

Similarly, a natural extension to the basic GC can be achieved by setting different reservation thresholds for each class of service. Pavlidou [89] analyzed an integrated voice/data cellular system using a two-dimensional Markov chain. Haung et al. [87] analyzed the *movable boundary* scheme with finite data buffering.

In the movable boundary scheme, voice and data traffic each have a dedicated set of the available channels. Once dedicated channels are occupied, voice and data calls will compete for the shared channels. Wu et al. [67], [70] considered a different approach in which voice and data calls first compete for the shared channels and then will use dedicated channels, which can be considered as a natural extension of GC. Interested readers are referred to [71] for a discussion on fixed and movable boundary schemes. A general discussion on bandwidth allocation schemes for voice/data integrated systems can be found in [72].

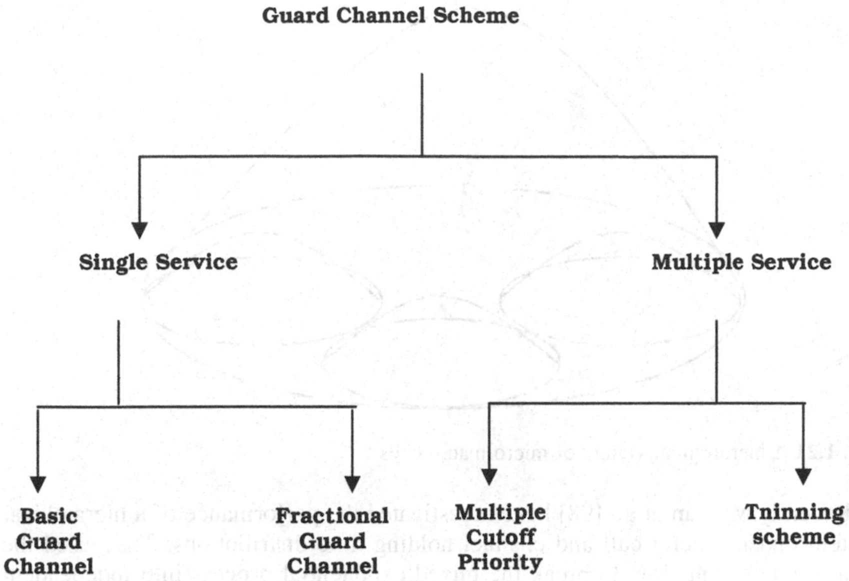

Guard Channel Scheme

```
                          Guard Channel Scheme
                                   |
                                   |
                 +-----------------+-----------------+
                 |                                   |
                 v                                   v
          Single Service                      Multiple Service
                 |                                   |
        +--------+--------+              +-----------+-----------+
        |                 |              |                       |
        v                 v              v                       v
     Basic           Fractional      Multiple                Tninning
     Guard             Guard          Cutoff                  scheme
    Channel           Channel        Priority
```

Fig. 1.20 Single service and multiple services guard channel schemes.

1.6.2 Hierarchical Schemes

As mentioned earlier, micro-pico-cell systems can improve spectrum efficiency better than macro cell systems because they can provide more spectrum resources per unit coverage area. However, micro-pico- cell systems are not cost effective in areas with low user population (due to base station cost) and areas with high user mobility (leading to a large number of handoffs).

As a consequence, hierarchical architectures [73–76] were proposed to take advantage of both macro cell and micro cell systems. Fig. 1.21 shows an example of a hierarchical cellular system. In this architecture, overlaid microcells cover high-traffic areas to enhance system capacity. Overlaying macro cells cover all of the area to provide general service in low-traffic areas and to provide channels for calls overflowing from the overlaid microcells. In particular, in a hierarchical system with an overflow scheme, it seems more significant to support guard channel for handoff protection and buffers for new and handoff calls in overlaying macro cells than to provide them in microcells [77]. In overflow schemes, when a call is rejected in a micro-cell, it is considered for admission by the macro-cell covering the micro-cell area.

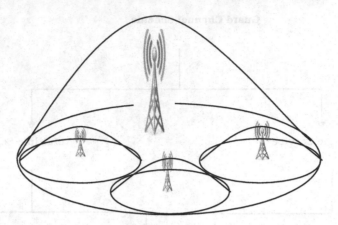

Fig. 1.21 A hierarchical system of micro/macro cells.

Recently Marsan et al. [98] have investigated the performance of a hierarchical system under general call and channel holding time distributions. They used the idea of equivalent flow to break the mixed exponential process into independent exponential processes, which can be then solved using classical Markov analysis.

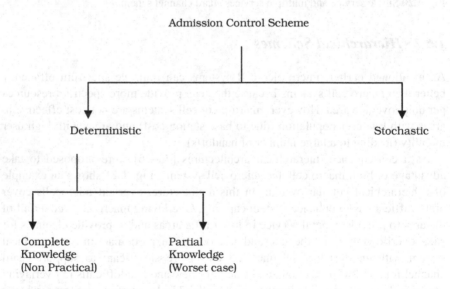

Fig. 1.22 Call admission control schemes.

1.6.3 Complete Knowledge Schemes

User mobility has an important impact in wireless networks. If the mobility pattern is partially [14] or completely [79] known at the admission time then the optimal decision can be made rather easily.

Many researchers believe that it is not possible in general to have such mobility information at admission time. Even for indoor environments complete knowledge is not available]. Nevertheless, such an imaginary perfect knowledge scheme is helpful for benchmarking purposes [79]. Fig. 1.22 depicts a classification of CAC schemes according to their knowledge about user mobility. Partial knowledge schemes must reserve resources in several cells [14] to provide deterministic guarantees; hence we call them *worse case* schemes.

In addition to CAC schemes assuming deterministic mobility information, there is a large body of research work addressing the probabilistic estimation and prediction of mobility information. Some of them are heuristic-based [28], [45], [80], [81], some others are based on geometrical modeling of user movements and street layouts [82], and some others are based on artificial intelligence techniques [83]. For instance, the distributed CACs introduced before are based on probabilistic mobility information.

1.6.4 Economic Schemes

Economic models are widely discussed as a means for traffic management and congestion control in provider's networks [84–86]. Through pricing, the network can send signals to users to change their behavior. It has been shown that for a given wireless network there exists a new call arrival rate, which can maximize the total utility of users [86]. Based on this, the admission control mechanism can adjust the price dynamically according to the current network load in order to prevent congestion inside the network.

In terms of economics, utility functions describe user's level of satisfaction with the perceived QoS; the higher the utility, the more satisfied the users. It is sometimes useful to view the utility functions as of money a user is willing to pay for certain QoS. As mentioned earlier, call blocking and dropping probabilities are the fundamental call-level QoS parameters in cellular networks. Let us define the QoS metric ϕ as a weighted sum of the call blocking and dropping probabilities as follows

$$\phi = \alpha p_b + \beta p_d \tag{1.21}$$

where α and β are constants that denote the penalty associated with blocking a new call or dropping an ongoing call respectively (with $\alpha > \beta$ to reflect the costly call dropping). Earlier we showed that pb and pd are functions of new call and handoff call arrival rates v and λ. Therefore

$$\phi = f(\lambda) \tag{1.22}$$

where f is a monotonic and non decreasing function of λ. Let us define U as the user utility function in terms of the QoS metric Φ, and let U = g(Φ), where g is a monotonic and non-increasing function of Φ. Therefore, the utility function U is maximized at $\Phi = 0$. Let λ^* denote the optimal arrival rate for which U is maximized. In [86], it has been shown that the sufficient condition for λ^* is that

$$\left.\frac{dU}{d\lambda}\right|_{\lambda=\lambda^*} = 0 \tag{1.23}$$

Using the optimal arrival rate λ^* obtained, we can characterize a pricing function to achieve the maximum utilization. Let p(t) denote the price charged to users at time t. Define H(t) as the percentage of users who will accept the price at time t, then

$$\lambda_{in}(t) = (\lambda(t) + v(t))H(t), \quad 0 \le H(t) \le 1 \tag{1.24}$$

where $\lambda_{in}(t)$ is the actual new call arrival rate at time t. H(t) must be designed in such a way that always

$$\lambda_{in}(t) \le \lambda^*, \tag{1.25}$$

and consequently

$$H(t) \le \min\left\{1, \frac{\lambda^*}{\lambda(t) + v(t)}\right\} \tag{1.26}$$

As mentioned before, pricing can influence the way the users use resources and is usually characterized by demand functions. A simple demand function can be characterized as follows [86]

$$D(t) = e^{-\left(\frac{p(t)}{p_0} - 1\right)^2}, p(t) \ge p_0 \tag{1.27}$$

where P_0 is the normal price. In fact, D(t) denotes the percentage of users that will accept the price p(t). In order to realize control function H(t) we should have H(t) = D(t). The price that should be set at time t to obtain the desired QoS can be expressed as

$$p(t) = p_0\left(1 + \sqrt{\max\left\{0, -\ln\frac{\lambda^*}{\lambda(t) + v(t)}\right\}}\right) \tag{1.28}$$

is worth noting that pricing-based control assumes that network users are sensitive and responsive to price changes. If this is not true for a particular network, e.g. noncommercial networks, then price-based control can not be applied.

1.7 Call Admission Control Schemes Based on Fuzzy Logic and Evolutionary Algorithms

In this section we describe call admission control schemes using different algorithm.

1.7.1 Call Admission Control Using Fuzzy Logic

Jun Ye et al [88] proposed a call admission control (CAC) scheme using fuzzy logic for the reverse link transmission in wideband code division multiple access (CDMA) cellular communications. The fuzzy CAC scheme first estimates the effective bandwidths of the call request from a mobile station (MS) and its mobility information, and then makes a decision to accept or reject the connection request based on the estimation and system resource availability. Numerical results are given to demonstrate the effectiveness of the proposed fuzzy CAC scheme in terms of new call blocking probability/handoff call dropping probability, outage probability, and resource utilization.

Y. H. Chen et al [89] proposed outage-based fuzzy call admission controller with multi-user detection (OFCAC-MUD) is proposed for wideband code division multiple access (WCDMA) systems. The OFCAC-MUD determines the new call admission based on the uplink signal-to-interference ratios from home and adjacent cells and system outage probabilities. The OFCAC-MUD possesses both the effective reasoning capability of a fuzzy logic system and the aggressive processing ability of MUD. Simulation results reveal that OFCAC-MUD without power control (PC) improves the system capacity by 70.5% as compared to an SIR-based CAC-RAKE with perfect PC. It also enhances the system capacity by 53.9% as compared to an OFCAC-RAKE with perfect PC, by 6.7% as compared to an SIR-based CAC-MUD without PC and by 12.9% as compared to an OFCAC-MUD with perfect PC, given the same outage probability requirements. Moreover, OFCAC-MUD can prevent the violation of outage probability requirements in the hotspot environment, which is hardly achieved by SIR-based CAC.

Chung-Ju et al. [90] proposed a neural fuzzy call admission and rate controller (NFARC) scheme for WCDMA cellular systems providing multirate services. The NFARC scheme can guarantee the quality of service (QoS) requirements and improve the utilization of the system. Simulation results show that the NFARC scheme achieves low forced termination probability and high system capacity even in the bursty traffic conditions. NFARC accepts users more than intelligent call admission controller (ICAC) by an amount of 45.35%.

In the present and next generation wireless networks, cellular system remains the major method of telecommunication infrastructure. Since the characteristic of the resource constraint, call admission control is required to address the limited resource problem in wireless network. The call dropping probability and call blocking probability are the major performance metrics for quality of service (QoS) in wireless network. Chenn-Jung Huang et al[91] proposed an adaptive call admission control and bandwidth reservation scheme using fuzzy logic control concept to reduce the forced termination probability of multimedia handoffs.

The authors adopt particle swarm optimization (PSO) technique to adjust the parameters of the membership functions in the proposed fuzzy logic systems. The simulation results show that the proposed scheme can achieve satisfactory performance when performance metrics are measured in terms of the forced termination probability for the handoffs, the call blocking probability for the new connections and bandwidth utilization.

1.7.2 Call Admission Control Using Genetic Algorithm(GA)

Shyamalie Thilakawardana and Rahim Tafazolli in [92] anticipated that a wide variety of data applications, ranging from WWW browsing to Email, and real time services like packetized voice and videoconference will be supported with varying levels of QoS. Therefore there is a need for packet and service scheduling schemes that effectively provide QoS guarantees and also are simple to implement. This paper describes a novel dynamic admission control and scheduling technique based on genetic algorithms, focusing on static and dynamic parameters of service classes. A performance comparison of this technique on a GPRS system is evaluated against data services and also a traffic mix comprising voice and data.

Sheng-Ling Wang et al [93] introduced an adaptive threshold-based Call Admission Control (CAC) scheme used in wireless/mobile network for multiclass services is proposed. In the scheme, each class's CAC thresholds are solved through establishing a reward penalty model which tries to maximize network's revenue in terms of each class's average new call arrival rate and average handoff call arrival rate, the reward or penalty when network accepts or rejects one class's call etc. To guarantee the real time running of CAC algorithm, an enhanced Genetic Algorithm is designed. Analyses show that the CAC thresholds indeed change adaptively with the average call arrival rate. The performance comparison between the proposed scheme and Mobile IP Reservation (MIR) scheme shows that with the increase of average call arrival rate, the average new Call Blocking Probability (CBP) and the average Handoff Dropping Probability (HDP) within 2000 simulation intervals of the proposed scheme are confined to lower levels, and they show approximate periodical trends of first rise and then decline. While these two performance metrics of MIR always increase. At last, the analysis shows the proposed scheme outperforms MIR in terms of network's revenue.

Shengling Wang et al [94] introduced a dynamic multi-threshold CAC scheme is proposed to serve multi-class service in a wireless/mobile network. The thresholds are renewed at the beginning of each time interval to react to the changing mobility rate and network load. To find suitable thresholds, a reward-penalty model is designed, which provides different priorities between different service classes and call types through different reward/penalty policies according to network load and average call arrival rate. To speed up the running time of CAC, an Optimized Genetic Algorithm (OGA) is presented whose components such as encoding, population initialization, fitness function and mutation etc., are all optimized in terms of the traits of the CAC problem. The simulation demonstrates that the proposed CAC scheme outperforms the similar schemes, which means the optimization is realized. Finally, the simulation shows the efficiency of OGA.

Bo Rong et al [95] proposed a mobile agent (MA)-based handoff architecture for the WMN, where each mesh client has an MA residing on its registered mesh router to handle the handoff signaling process. To guarantee quality of service (QoS) and achieve differentiated priorities during the handoff, they develop a proportional threshold structured optimal effective bandwidth (PTOEB) policy for call admission control (CAC) on the mesh router, as well as a genetic algorithm (GA)-based approximation approach for the heuristic solution. The simulation study shows that the proposed CAC scheme can obtain a satisfying tradeoff between differentiated priorities and the statistical effective bandwidth in a WMN handoff environment.

1.7.3 Call Admission Control Using Neural Network (NN)

Conventional methods for developing CAC algorithms are based on mathematical and/or simulation modeling. These methods require making assumptions about the traffic processes. QoS prediction is then done using queuing models to reflect the buffering and transmission behavior. Since this approach can quickly become analytically involved, simplifying assumptions need to be made. For example, it is common to assume that traffic sources are Markovian, or stationary, or that cell arrival patterns depend upon some parametric models like Markov Modulated Poisson Processes (MMPP).

It has recently become known that high-speed network traffic is more complex, and that none of these assumptions is safe. Exact solutions based on analytical methods exist only for restricted traffic and system models. QoS estimation through analysis can also be inaccurate because declared and actual traffic parameters frequently differ.

Clearly results derived from analyses based on such assumptions have limited applicability and will generate inaccurate QoS estimates. To compensate, such CAC schemes force themselves to err on the side of being conservative and thus typically over allocate resources. This leads to inefficiency.

The easiest method for CAC is to accept a call if there is enough available bandwidth to allocate to the call its peak rate. This is the most inefficient method since it entirely ignores statistical multiplexing. The most well known analytical result for CAC is the equivalent bandwidth method [100,101]. This method provides a simple formula to compute the amount of bandwidth needed to meet a call's loss requirement, given its peak rate, mean rate, and average burst duration. The equivalent bandwidth yields a bandwidth that lies between the call's peak and mean rates. This method is exact only asymptotically, as the buffer size approaches infinity and the cell loss probability approaches zero. Although far superior to a CAC scheme that allocates peak bandwidth, the equivalent bandwidth method still over allocates bandwidth in most cases.

Neural networks are attractive for solving CAC problems because they are a class of approximators that are well suited for learning nonlinear functions. A neural network represents a multiple-input multiple-output nonlinear mapping. A NN can learn this mapping from a set of sample data. Feed-forward neural networks can approximate any piecewise-continuous function with arbitrary accuracy, given enough hidden neurons [102].

Some advantages of using neural networks for CAC are:

1. Neural networks do not require an accurate mathematical model of either the traffic or the system. No assumptions need to be made since the neural network is trained on observed data. Not assuming a specific traffic behavior a priori is a preferable approach because multimedia traffic is not well understood and continuously changing. NNs are also not affected by mistakes in declared traffic descriptors. These features allow a NN to yield more accurate QoS estimation which leads to greater efficiency and robustness.

2. When NNs are properly trained, they can generalize and extrapolate additional details of the function mapping the inputs to outputs. If the training set if sufficiently large, a NN will generalize accurately and will produce accurate outputs for inputs not in the training set. This also contributes to robustness of the CAC scheme.

3. NNs are adaptable since they can be retrained in real-time using the latest measurements

1.7.3.1   Implement CAC with Neural Networks

Neural networks can be used to solve the CAC problem and all its variants. First we describe a model using a NN for the single-buffer-per-link case, with a FIFO (first-in first-out) buffer. The NN contains n inputs, denoted $\{u_0, u_1, \dots u_n\}$ and a single output $Y = f(u)$, where f(u) denotes the transfer function from inputs to outputs. The output is the QoS estimate. This can be the estimate for the amount of bandwidth needed on the link to carry all the calls whose traffic enters the FIFO buffer, or the buffer delay, or the buffer's loss rate. This version of this model, with a single output representing a particular QoS estimate, only supports a service that makes guarantees on one QoS parameter. For services that provide guarantees on m QoS parameters, a version of this model with m outputs, one per QoS parameter, could be used.

In another version of this model, a single output is used that takes on binary values that represent accept or reject decision. In this version, particular QoS estimates are internal to the NN. Such a NN is called a classifier. With this choice for the output, the NN can be used to represent any definition of a feasible stream, including definitions that involve multiple QoS constraints (e.g., delay and CLR).

In the case of a single output representing a QoS estimate, a call is admitted to the network if the QoS estimate for the new candidate aggregate stream is below the most stringent QoS requirement (i.e., the decision threshold) for all calls in the stream. In the case of multiple outputs, each QoS estimate needs to be compared the relevant decision threshold, before a call is accepted. In the case of a single binary output (e.g., trained to learn 0 for accept and 1 for reject), the output is compared to a threshold such as 1/2. The choice of output influences by which CAC problem is being modeled. Consider the case in which the output is a queue related parameter, such as delay or loss. When FIFO queuing is used, there is only a single loss rate or distribution of delay associated with that queue.

The NN predicts the QoS of the aggregate stream of superposed traffic sources. When many sessions that have different QoS requirements are multiplexed together, the switch ensures the most stringent of all the delay and/or loss requirements. (Hence, some sessions will experience better QoS that they requested.) Thus this model can support multiple traffic classes, but they will all receive the same QoS. If a scheduling mechanism (e.g., earliest deadline first) is used to prioritize among traffic classes, then multiple loss rates (one for each class) could be computed for a single buffer. A NN used for this scenario would require multiple outputs, one for each class.

We now consider a second model for the multiple-buffers-per-link problem. In this case, there is at least one QoS estimate per buffer. If there are b buffers, and x QoS parameters per buffer, then this NN model could have b_x outputs. Alternatively, a single binary output (representing an accept or reject decision for a given call) could still be used in this case, since all call types and services share a single transmission link.

A third way of modeling CAC problems is to use a modular design in which multiple NN units are organized in a hierarchical fashion. Modularity is a means to solve a complex computational task by decomposing it into simpler subtasks, and then combining the individual solutions. This approach is attractive if the functional relationship between the neural network inputs and outputs is very complex, and if parts of this function can naturally be separated. Consequently, the modules of the network tend to specialize by learning different regions of the input space. The decomposition should be structured so as to facilitate this.

An example, based on the work in [115], is given in Figure 1.23. There is a bank of NN units in level 1 of this model. Each NN unit is associated with a single traffic class, and its inputs correspond to descriptors for that class. Let's assume that one virtual path is assigned for each traffic class. The output gives the bandwidth estimate for all the calls in that class, i.e., for the VP assigned to that class. The NN unit at level 2 takes as inputs the bandwidth per class for each VP and outputs the link bandwidth needed to support the VPs.

The attraction of this model is that it naturally separates out the two levels of statistical multiplexing that occur in a switch that supports VPs. The NNs in level 1 determine the amount of multiplexing gain for mixing calls from the same class onto a single VP, while the NN in level 2 determines the amount of multiplexing gain for mixing VPs onto a link. Another example of a modular NN design can be found in [114].

Modular networks offer several advantages over a single neural network [102]. First, the training is faster, which allows the NN units to be more adaptable. Second, the representation of input data developed by a modular network tends to be easier to understand than in the case of an ordinary multilayer NN. Third, this type of design should lead to more accurate estimates since the input-output mapping that each NN unit has to learn is simpler. It has been proven [96] that the number of input-output patterns that a NN can deterministically learn is equal to twice the number of its weights. Fourth, a useful feature of a modular approach is that it also provides a better t to a discontinuous input-output mapping [102].

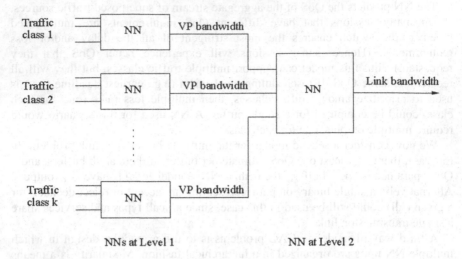

Fig. 1.23 Sample Modular NN Design for CAC

Architecture: A common architecture for all the models described above is a 3-layer feed forward neural network, with a layer of inputs, a single hidden layer of neurons, and an output layer. We establish the following notation for the neural network elements.

For a given set of inputs $u_1,......,u_n$, the k-th output of the NN is given by

$$Y_k = g^{out}(\sum_{j=1}^{J} W_{jk}V_j + b_k^{out}) \tag{1.29}$$

where W_{jk} is the connection weight from the j-th hidden neuron to the k-th output, and b_k^{out} is the bias for the k-th output. $V_j, j = 1,...J, V\, j$, is the output of the jth hidden neuron, and is given by

$$V_j = g(\sum_{i=1}^{l} w_{i,j}u_i + b_j) \tag{1.30}$$

The w_{ij} are the weights from the inputs to the hidden neurons, and b j are the biases for the hidden neurons. The functions gout and $g(x)$ are the activation functions for the output layer and the hidden neurons, respectively. These functions are typically either a linear function or a sigmoid function, such as the logistic function $_1 = (1 + \exp(av))$ or the hyperbolic tangent function $\tanh(v = 2) = (1 + \exp(v))$. In the case of a single output, k = 1, the subscript k can be dropped.

The key to a good NN design lies in the particular choice of inputs, outputs and training technique. Selecting a pair of inputs and outputs for which the desired output is not determinable from the input will clearly not yield an effective CAC algorithm.

1.7.3.2 NN Inputs

The NN inputs can be any input that helps the NN to predict the QoS of an aggregate traffic stream. The NN inputs typically include either traffic descriptors or system state parameters or some combination of both. In this section, we now expand the discussion of these types of inputs. The advantage of having the users supply traffic descriptors is that the network need not spend any resources or time to measure the traffic.

However the disadvantage comes from the fact that there is usually a difference between the declared traffic parameters and the actual traffic parameters since most applications today do not understand well the traffic they generate. When the number of connections in a network becomes large, the difference between the declared and actual traffic parameters can be quite large. The advantages and disadvantages of using measurements as NN inputs are exactly the reverse of the advantages and disadvantages of the user supplied approach.

Examples of system state parameters that can be used as inputs include buffer level, the number of existing calls for each traffic class, and buffer loss rates. It is desirable for the neural network inputs to have the following properties:

Capture key elements of traffic behavior that influence queuing. Many researchers believe that traffic descriptors that capture the correlation and burstiness properties of a traffic stream will be successful for CAC. In order to avoid over allocating or under-allocating resources, it is necessary estimating QoS well, which in turn requires proper traffic and system characterization.

When the NN input vector is additive, the current input vector can be updated efficiently by simply adding the traffic descriptors of the new call to that of the aggregate call (the current input vector). This additive property of traffic descriptors greatly speeds up the decision process of accepting or rejecting a call.

Support a large number of traffic classes. This will make the algorithm robust. Keep the number of inputs reasonably small. This will make the algorithm practical since the forward calculation speed is proportional to the number of weights. We now give examples of NN inputs that have been proposed in recent research efforts.

1.7.3.3 Number of Calls per Traffic Class

In this case, the NN input is a vector $s = (s\ k)$ whose k^{th} component gives the number of calls in the stream that belong to the kth traffic class. This input has been used by [105, 106, 114]. In [110], they use this input coupled with the link load level. We refer to this particular input as the call vector in subsequent sections of this chapter. The advantage of this approach is that the user need not supply any traffic descriptors at all, and the decision boundary is determinable from this input. The disadvantage of this approach is that it does not scale well in the number of traffic classes. There could be a very large number of traffic classes in any network. It has been suggested [106, 114] that a practical number of classes is less than 100.

1.7.3.4   Counts of Arrivals

In [115] they used on-line traffic measurements for the NN inputs. In each interval q, the number of cell arrivals, N S(q), is counted for each stream S. If one keeps track of the arrival process over consecutive intervals, and also provides this data to the NN, then the NN can be trained to capture the correlations that exist among cell arrivals. This approach requires a careful choice of the measurement interval. The advantages of this input are that this information entirely characterizes the input stream and that the user need not supply any traffic descriptors. The disadvantage of this choice for NN inputs is that is does not scale in the number of calls supported simultaneously. This choice of inputs is not considered very practical since the measurement intervals typically need to be very large.

1.7.3.5   Variance of Counts

In [111], we used the variance of counts (VOCs) as NN inputs. To calculate or measure VOCs, time is divided into intervals of equal length. As above, N S(q) denotes the number of cells or packets arriving in interval q for stream S. Let Ns (q; h) denote the number of cells arriving in the interval consisting of intervals q through h. Let S denote the mean of Ns(q). The variance of counts for an interval of length m is defined by

$$VOC_s(m) = \frac{\mathrm{var}\{N_s(q+1, q+m)\}}{m} \tag{1.31}$$

where q is an arbitrary time slot. For a given stream S, the NN inputs are scaled versions of S and $VOC_s(m)$ for m = 1; 2; 4; ::::; 2 M , where M + 1 is the number of VOCs used. To limit the number of NN inputs while considering a representative set of VOCs, we used VOCs over intervals of exponentially increasing length.

The VOC traffic descriptor is an not normalized IDC as described in [103], that is, VOC S(m) = IDC S(m) S. Not normalized IDCs are preferable because then the VOCs are additive, i.e., if S is the sum of statistically independent streams S $_i$, then VOC S(m) is the sum of VOC Si(m) over the Si .

The advantages of these NN inputs include: (i) VOCs characterize all second-order statistics of the stream, (ii) they are additive, (iii) moments of interval counts have been shown to accurately predict queuing delay for some models [103], and (iv) this method is independent of the number of traffic classes.

It is known [103] that IDC S(m) converges to var(X) = $E^2(X)$, the squared coefficient of variation of the inter-arrival time X, and so VOC S(m) also converges to a constant if the inter-arrival time has finite variance. One can thus choose the number M + 1 of VOCs to be large enough so that $VOC_s(2M)$ is close to the limit for most streams. This limits the number of NN inputs to M + 2, which is typically much smaller than the number of traffic classes as used in

These NN inputs can be measured by either the user or the network. They could be calculated by the user when the user's traffic is a bursty on-off process. VOCs can be prior computed for a set of traffic classes and stored in a table, in which case a user only needs to indicate a traffic class in the call request.

1.7.3.6 Power Spectral Density Parameters

Another type of input that can capture correlation and burstiness properties of traffic streams is the power-spectral-density (PSD) function in the frequency domain. Using the PSD as inputs was first proposed in [87]. All of the above inputs discussed focus on the time-domain. The PSD is the Fourier transform of the autocorrelation function of the input process. A traffic source can be characterized by a PSD can be described by three parameters (u;v;w): the DC component (u), the half-power bandwidth (v), and the average power (w). As u increases, the traffic loads increases; as v decreases, the input power in the low frequency band increases; and as w increases, the variance of the input rate increases.

There are two advantages to this approach. First, it has been shown in [107] that the low frequency band of the input PSD has a dominant impact on the queuing performance while the high frequency band can usually be neglected. This is because the low frequency component of the PSD contains the correlation component. Larger the low frequency component becomes, the burstier the traffic source. Second, since the PSD has the additive property, so do these three parameters.

In [97], the CAC controller is designed so that the user can input three simple parameters: its peak rate, mean rate and peak cell rate duration. The controller applies the Fast Fourier Transform (FFT) to these inputs and outputs the three PSD parameters (u; v; w) which are in turn fed as inputs to a NN.

1.7.3.7 Entropy

Entropy has been proposed as a tra c descriptor in [99, 112]. The entropy of traffic streams is attractive as a descriptor because it can capture the behaviour of correlations over many time scales. The entropy has been used as an input for CAC in [99, 112] but it has not yet been tried in a CAC algorithm based on neural networks.

1.7.3.8 NN Outputs

A NN output variable can represent any of the following:

1. **Accept/Reject Decision.** With this output, the NN has to learn the boundary between the feasible and infeasible performance regions for a given input space.
2. **Loss Rate.** In this case, the NN predicts the average buffer overflow rate. Since the loss rate can have an exponentially wide range, from 10-1 to 10-9 it is common to use log(loss) instead.
3. **Delay.** In this case, the NN predict the average buffer delay, or average buffer occupancy.
4. **Jitter.** In this case, the NN predicts the variation of the buffer delay. 5. **Bandwidth.** In this case, the NN predicts the amount of bandwidth needed to achieve a specific QoS level for the given input stream.

5. **P^{th} percentile delay.** In this case, the NN predicts the value D such that the probability that a cell or packet experiences a delay less than or equal to D is p%. The percentile is typically chosen to be around 90%. In this approach, all the calls need to have the same percentile requirement.

6. **Probability distribution of delay.** In this case, the NN output represents the probability that the delay experienced will be less than the requested delay (which can be one of the NN inputs), conditioned on the buffer status (represented by the other NN inputs). This method, proposed in [113], works well when the probability is conditioned with respect to the number of active connections. This method allows calls to have different delay requirements.

1.7.3.9 Compression of Neural Network Inputs

In Section 1.7.3.5, we discussed the mapping of the call vector to a vector of parameters related to second order statistics of the aggregate traffic process (based on VOCs or the PSD). Such a mapping can be considered a compression of the call vector (whose dimension is the number of traffic classes, which can be in the hundreds) to a smaller vector whose dimension is independent of the number of traffic classes. One benefit of reducing the number of NN inputs is that the NN output can be computed in less time, thus allowing more calls to be processed per second. The best compression is one that maps the call vector to the fewest parameters without reducing performance significantly.

In this section, we present a method for finding a linear compression of the call vector to I parameters (where I is any positive integer) that is optimal in the sense of minimizing the output error function. Linear compression is chosen so that the compressed parameters are additive. A similar method was presented in [111] to compress vectors of VOCs, where it was shown that compression to three parameters did not result in a significant reduction of performance. An advantage of compressing call vectors instead of VOCs is that the statistics of calls need not be known or measured.

Although we focus on the call vector in this section, the method can be applied to any choice of NN inputs that are additive. The compression method involves adding a new hidden layer to the NN, between the input layer and the current hidden layer, as shown in Fig. 1.24. The neurons in the new hidden layer have linear activation functions, and there is one such neuron for each compressed parameter.

For this design, the previous equation described in architecture is replaced by the following two equations

$$V_j = g\left(\sum_{i=1}^{I} w_{ij} + b_j\right) \qquad (1.32)$$

$$v_i = \sum_{l=1}^{L} \alpha_{li} s_l \qquad (1.33)$$

The outputs v_i of the hidden linear neurons form the compressed vector corresponding to the call vector s, and the weights l_i from the lth input to the first hidden layer form the compressed vector corresponding to the lth traffic class. Above equation can be expressed in vector form as

$$v = \sum_{l=1}^{L} \alpha_l s_l \qquad (1.34)$$

where v is the compressed vector for call vector s and l is the compressed vector for a single call of traffic class 1.

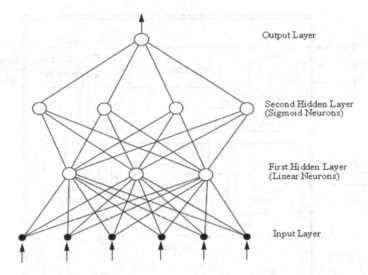

Output Layer

Second Hidden Layer
(Sigmoid Neurons)

First Hidden Layer
(Linear Neurons)

Input Layer

Fig. 1.24 NN with Additional Hidden Layer for Compressing Input

This method was motivated by the well-known technique of using a hidden layer of linear neurons for image or data compression [102, 104, 108]. Assuming that NN weights (including the l_i) are found that minimize the output error function, the compressed parameters v i are optimal by definition. Once this NN is trained and thus the matrix $\{\alpha_{li}\}$ is obtained, the original input layer is no longer required. The first hidden layer becomes the new input layer; i.e., the compressed parameters v_i are now used as the NN inputs.

Since the compressed parameters are additive, when a new call of class 1 arrives, the compressed vector for the new call vector is updated simply by adding the compressed vector 1 for the new call. To perform this operation quickly, the compressed vector 1 corresponding to each traffic class 1 can be stored in a look-up table.

Consider the special case in which only a single compressed parameter is used. Thus, a single parameter α_l is computed for each traffic class 1, and the compressed parameter for a given call vector is equal to the sum of α_l over all calls represented by the call vector. Moreover, it is possible to scale the α_l so that a given call vector falls into the call acceptance region (learned by the neural network) if and only if this sum is less than the link bandwidth. Therefore, compression to a single parameter corresponds to learning the equivalent bandwidth for each traffic class.

1.7.3.10   Design of CAC Controller

Fig. 1.25 depicts a method of incorporating a NN into the design of a CAC algorithm.

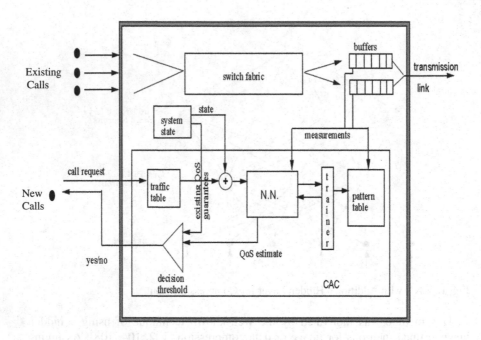

Fig. 1.25 Design of CAC Controller using a Neural Network

This figure is general in that it illustrates potential CAC inputs, and not required inputs. The CAC inputs can come from the user, the system state or measurements, or some combination thereof. A traffic table can be used to convert user supplied traffic descriptors into VOCs or PSDs, if needed. If both the system state and the user's traffic descriptors (or the mapped version of the user's descriptors) are used, we assume they have the additive property (indicated by the adder in Fig. 1.25). The system state is updated each time new connections are established or torn down. The pattern table and trainer are only present when the system is designed for on-line training.

1.7.3.11 Off-Line Training

The feed forward NNs described above can be trained using standard back propagation algorithms and their variations. In this section we brie y discuss methods for generating the training set (the set of input-output patterns) used for back propagation and present a non-standard error function that helps to achieve the asymmetric goal of the CAC problem. Methods for the more difficult problem of online training are presented in the next section. When a NN is trained offline for CAC, the training set consists of a large number (typically 1000 or more) of input-output patterns (u^k, y_k) that are usually generated by simulating the traffic and queuing processes. The traffic can be modeled as an on-o Markov chain, a Markov-modulated Poisson process, or a self-similar traffic process. More complex traffic models or traces of traffic from actual network can also be used. A fixed number of traffic classes can be defined by specifying the parameters for each class, or an infinite number of classes can be obtained by allowing any choice of parameters within some range.

Aggregate traffic streams must be generated that cover all regions of the space of input patterns u^k. If the number of traffic classes and the maximum number of calls per class are small, it may be possible to use a training set that includes every possible call state. Otherwise, one way to generate an aggregate stream is to rst randomly select the number of calls in the stream (between 1 and the maximum possible number of calls), and then randomly select the traffic class for each call. If the input pattern u k is the call state, then it is known immediately. If some other choice is used for u^k (e.g., VOCs), then u^k must be obtained using analysis or simulation. For each aggregate stream, a simulation can be run to determine the resulting performance measure y_k. Each simulation should be run long enough to obtain an accurate estimate of the performance measure. For example, 10^8 packets would need to be observed to accurately estimate a loss rate of 10^{-6}. If such long simulations are not feasible, the virtual output buffer method discussed in the next section can be used to improve the estimate through extrapolation.

Once the training set is generated, the NN can be trained using a version of back propagation. A commonly used error function for training is

$$J = \sum_{k=1}^{K} \left| f(u^k) - y_k \right|^N \tag{1.35}$$

where, usually $N = 2$ (equivalent to the mean squared error).

Recall that the CAC objective is asymmetric: to accept as many feasible input patterns as possible while rejecting all infeasible input patterns. After the NN is trained, the decision threshold can be adjusted so that all infeasible input patterns in the training set are rejected. That is, assuming that a NN output greater than the decision threshold corresponds to a reject decision, the threshold is chosen so that it is slightly less than the smallest NN output for an infeasible input pattern in the training set.

In order to accept as many feasible streams as possible, the maximum error over all infeasible training patterns u should be minimized, so that the decision threshold can be chosen as large as possible without accepting any infeasible patterns. One way to achieve this objective is to use the following asymmetric error function

$$J = \sum_{feasible\,k} \left| f(u^k) - y_k \right|^2 + \sum_{infeasible\,k} \left| f(u^k) - y_k \right|^N \qquad 1.36$$

where N > 2. For large N, minimizing the second sum will tend to minimize the maximum error $\left| f(u^k) - y_k \right|$ over all infeasible patterns.

1.7.3.12   Online Training

In online training, the NN is trained continuously or frequently, based on actual measurements obtained while the NN is being used for CAC. Online training is useful if the CAC needs to adapt to changing network characteristics or new traffic classes or to fine tune an offline-trained NN using more accurate measurements.

Online training is more difficult than offline training in part because the call state changes frequently. Thus, the performance measurement (NN output) obtained for a given training pattern is based on fewer packets (or cells) and is therefore subject to more statistical variability (i.e., noise). In online training, unlike offline training, one cannot x the call state and observe the performance of a large number of packets in order to obtain a good estimate of the resulting loss rate or average delay. This difficulty can be reduced by exploiting the ability of NNs to learn the average of several different measurements associated with the same call state, as explained below.

However, online training can be slow because a large number of packets must be observed before a small loss rate can be estimated with any accuracy. For example, more than a billion packets must be observed to accurately estimate a loss rate on the order of 107. In this case, a packet loss can be considered a rare event. The online training can be made faster by using the ability of a neural network to extrapolate from estimates of measures that are based on more common events. For example, in the virtual output buffer method, discussed below, the NN learns the loss rate that would occur if the packets were fed into imaginary queues with smaller service rates than the actual queue, and extrapolates that knowledge to improve the estimate of the actual loss rate.

For a NN CAC to be adaptive, old measurements for a given input pattern must eventually be "forgotten" and replaced by new measurements for similar input patterns. There is a tradeoff between adaptability and accuracy: If the NN is trained using past measurements made over a large time window, it can achieve good accuracy but will not adapt quickly to network changes. If a small time window is used (so that the NN quickly forgets past measurements), faster adaptation is achieved at the cost of less accuracy.

Another reason online training is more difficult than offline training is that input patterns that are marginally unacceptable occur rarely or never, assuming the CAC is performing well. If the NN remembers that these patterns are unacceptable, it may not be able to adapt to changing network conditions that cause these patterns to become acceptable. If the NN eventually "forgets" these patterns, it will start to accept infeasible call vectors and thus perform poorly for some period of time while it is relearning the decision boundary. Such behavior should be avoided, since it is more important to reject a given infeasible call vector than to accept a given feasible call vector, i.e., the NN should achieve "safe-side" control. One way to help achieve this goal to start with a NN CAC that has been trained offline to perform a conservative version of CAC, such as one based on peak rate or equivalent bandwidth, and then use online training with a slow learning rate so that the CAC gradually learns to accept more calls. The virtual buffer method also helps to achieve safe-side control by learning more quickly that the call vector is approaching the decision boundary.

Other problems that will be addressed in this section include how to decide which patterns to store, given a bounded storage capacity, and summarizing different measurements for the same input pattern, in order to reduce the training set and thus reduce the time required to train the NN. In the following subsections, we assume that the NN input is the call vector and that the NN has a single output. However, the methods are applicable to other choices for the NN input and can easily be extended to multiple NN outputs.

1.7.3.13 Procedures for Online Training

In online training, we cannot assume that the queuing system reaches equilibrium between changes in the call vector. Therefore, the neural network can only learn, for each possible state of the call vector, the average performance of packets that are admitted when the call vector is in that state. A training pattern will therefore consist of a pair (s^k, y_k) and a size m_k where y_k is a measurement of the average performance of m_k packets that were admitted when the call vector was s^k. Since the call vector can change at arbitrary times, the intervals over which these measurements are performed need not have equal length. In an extreme case, each measurement can correspond to a single packet.

The simplest method for online training is to perform one step of gradient descent (back-propagation) per measurement, in the same order that the measurements are observed. However, this method does not provide the benefit of storing a large window of past measurements, including measurements for call vectors that rarely occur, and using each pattern repeatedly for training, so that better convergence is achieved and past measurements are not quickly forgotten.

Therefore, training patterns (each consisting of a call state and a corresponding measurement) should be stored in a pattern table. (See Fig 1.25) The time window over which past measurements are stored can be selected depending on the desired degree of adaptability. If the network characteristics are not expected to change and good estimation accuracy is desired, then a very large time window can be used. In addition, the total number of patterns that are stored should be limited, either because of storage limitations or to limit the time required to train the NN. To achieve this, a circular buffer can be used, so that the newest pattern replaces the oldest pattern.

1.8 Conclusion

Due to the unique characteristics of mobile cellular networks, mainly mobility and limited resources, the wireless resource management problem has received tremendous attention. As a result, a large body of work has been done extending earlier work in fixed networks as well as introducing new techniques. A large portion of this research has been in the area of call admission control. In this paper, we have provided a survey of the major call admission control approaches and related issues for designing efficient schemes. A broad and detailed categorization of the existing CAC schemes was presented. For each category, we explained the main idea and described the proposed approaches for realizing it and identified their distinguishing features.

We have compared the various schemes based on some of the most important criteria including efficiency, complexity, overhead, adaptively and stability. We believe that this article, which is the first comprehensive survey on this subject, can help other researchers in identifying challenges and new research directions in the area of call admission control for cellular networks.

References

1. Viterbi, A.: Principles of Spread Spectrum Communication. Addison–Wesley (1995)
2. TIA/EIA/IS–95, Mobile Station–Base Station Compatibility Standard for Dual–Mode Wideband Spread Spectrum Cellular System. Telecommunication Industry Association (May 1995)
3. Cooper, G.R., Nettleton, R.W.: A Spread–Spectrum Technique for High–Capacity Mobile Communications. IEEE Transactions on Vehicular Technology VT–27, 264–275 (1978)
4. Rappaport, T.S.: Wireless communications. Prentice Hall (1996)
5. Gilhousen, K.S., Jacobs, I.M., Padovani, R., Viterbi, A.J., Weaver Jr., L.A., Wheatly III, C.E.: On the capacity of a cellular CDMA system. IEEE Transactions on Vehicular Technology 40, 303–312 (1991)
6. Viterbi, A.J.: Wireless Digital Communication: A View Based on Three Lessons Learned. IEEE Communications Magazine 29, 33–36 (1991)
7. Jung, P., Baier, P.W., Steil, A.: Advantages of CDMA and Spread Spectrum Techniques over FDMA and TDMA in Cellular Mobile Applications. IEEE Transactions on Vehicular Technology 42, 357–364 (1993)
8. Corazza, G.E., De Maio, G., Vatalaro, F.: CDMA Cellular Systems Performance with Fading, Shadowing, and Imperfect Power Control. IEEE Transactions on Vehicular Technology 47, 450–459 (1998)
9. Wong, D., Lim, T.J.: Soft handoffs in CDMA mobile systems. IEEE Personal Communications Magazine 4(6), 6–17 (1997)
10. Prakash, R., Veeravalli, V.V.: Locally optimal soft handoff algorithms. IEEE Transactions on Vehicular Technology 52(2), 231–260 (2003)
11. Lin, Y.-B., Pang, A.-C.: Comparing soft and hard handoffs. IEEE Transactions on Vehicular Technology 49(3), 792–798 (2000)
12. Ramjee, R., Towsley, D., Nagarajan, R.: On optimal call admission control in cellular networks. Wireless Networks 3(1), 29–41 (1997)

13. Valko, A.G., Campbell, A.T.: An efficiency limit of cellular mobile systems. Computer Communications Journal 23(5-6), 441–451 (2000)
14. Talukdar, A.K., Badrinath, B., Acharya, A.: Integrated services packet networks with mobile hosts: Architecture and performance. Wireless Networks 5(2), 111–124 (1999)
15. Lu, S., Bharghavan, V.: Adaptive resource management algorithms for indoor mobile computing environments. In: Proc. ACM SIGCOMM 1996, Palo Alto, USA, pp. 231–242 (August 1996)
16. Hong, D., Rappaport, S.S.: Traffic model and performance analysis for cellular mobile radio telephone systems with prioritized and nonprioritized handoff procedures. IEEE Transactions on Vehicular Technology 35(3), 77–92 (1999)
17. Katzela, I., Naghshineh, M.: Channel assignment schemes for cellular mobile telecommunication systems: A comprehensive survey. IEEE Personal Communications Magazine 3(3), 10–31 (1996)
18. Chang, C.-J., Huang, P.-C., Su, T.-T.: A channel borrowing scheme in a cellular radio system with guard channels and finite queues. In: Proc. IEEE ICC 1996, Dallas, USA, vol. 2, pp. 1168–1172 (June 1996)
19. Wu, X., Yeung, K.L.: Efficient channel borrowing strategy for multimedia wireless networks. In: Proc. IEEE GLOBECOM 1998, Sydney, Australia, vol. 1, pp. 126–131 (November 1998)
20. Chu, T.-P., Rappaport, S.S.: Generalized fixed channel assignment in microcellular communication systems. IEEE Transactions on Vehicular Technology 43(3), 713–721 (1994)
21. Chang, C.-J., Su, T.-T., Chiang, Y.-Y.: Analysis of a cutoff priority cellular radio system with finite queueing and reneging/dropping. IEEE/ACM Transactions on Networking 2(2), 166–175 (1994)
22. Li, W., Chao, X.: Modeling and performance evaluation of a cellular mobile network. IEEE/ACM Transactions on Networking 12(1), 131–145 (2004)
23. Li, B., Chanson, S., Lin, C.: Analysis of a hybrid cutoff priority scheme for multiple classes of traffic in multimedia wireless networks. Wireless Networks 4(4), 279–290 (1998)
24. Lin, P., Lin, Y.-B.: Channel allocation for GPRS. IEEE Transactions on Vehicular Technology 50(2), 375–384 (2001)
25. Fang, Y., Zhang, Y.: Call admission control schemes and performance analysis in wireless mobile networks. IEEE Transactions on Vehicular Technology 51(2), 371–382 (2002)
26. Moorman, J.R., Lockwood, J.W.: Wireless call admission control using threshold access sharing. In: Proc. IEEE GLOBECOM 2001, San Antonio, USA, vol. 6, pp. 3698–3703 (November 2001)
27. Levine, D., Akyildiz, I., Naghshineh, M.: A resource estimation and call admission algorithm for wireless multimedia networks using the shadow cluster concept. IEEE/ACM Transactions on Networking 5(1), 1–12 (1997)
28. Choi, S., Shin, K.G.: Predictive and adaptive bandwidth reservation for handoffs in QoS-sensitive cellular networks. In: Proc. ACM SIGCOMM 1998, Vancouver, Canada, vol. 27, pp. 155–166 (October 1998)
29. Epstein, B.M., Schwartz, M.: Predictive QoS-based admission control for multiclass traffic in cellular wirelessnetworks. IEEE Journal on Selected Areas in Communications 18(3), 523–534 (2000)

30. Wu, S., Wong, K.Y.M., Li, B.: A dynamic call admission policy with precision QoS guarantee using stochastic control for mobile wireless networks. IEEE/ACM Transactions on Networking 10(2), 257–271 (2002)
31. Peha, J.M., Sutivong, A.: Admission control algorithms for cellular systems. Wireless Networks 7(2), 117–125 (2001)
32. Epstein, B., Schwartz, M.: Reservation strategies for multi-media traffic in a wireless environment. In: Proc. IEEE VTC 1995, Chicago, USA, vol. 1, pp. 165–169 (July 1995)
33. Box, G.E., Jenkins, G.M.: Time Series Analysis: Forecasting and Control, 2nd edn. Holden-Day (1976)
34. Zhang, T., Berg, E., Chennikara, J., Agrawal, P., Chen, J.C., Kodama, T.: Local predictive resource reservation for handoff in multimedia wireless IP networks. IEEE Journal on Selected Areas in Communications 19(10), 1931–1941 (2001)
35. Hosking, J.R.M.: Fractional differencing. Biometrika 83(1), 165–176 (1981)
36. Brockwell, P.J., Davis, R.A.: Time Series: Theory and Methods, 2nd edn. Springer (1991)
37. Gripenberg, G., Norros, I.: On the prediction of fractional brownian motion. Journal of Applied Probability 33, 400–410 (1996)
38. Norros, I.: On the use of fractional brownian motion in the theory of connectionless networks. IEEE Journal on Selected Areas in Communications 13(6), 953–962 (1995)
39. Leland, W.E., Taque, M., Willinger, W., Wilson, D.: On the self-similar nature of Ethernet traffic (extended version). IEEE/ACM Transactions on Networking 2(1), 1–15 (1994)
40. Crovella, M.E., Bestavros, A.: Self-similarity in world wide web traffic: Evidence and possible causes. IEEE/ACM Transactions on Networking 5(6), 835–846 (1997)
41. Beran, J., et al.: Long-range dependence in variable-bit-rate video traffic. IEEE Transactions on Communications 43(2), 1566–1579 (1995)
42. Shu, Y., Jin, Z., Wang, J., Yang, O.W.: Prediction-based admission control using FARIMA models. In: Proc. IEEE ICC 2000, New Orleans, USA, vol. 3, pp. 1325–1329 (June 2000)
43. Shu, Y., et al.: Traffic prediction using FARIMA models. In: Proc. IEEE ICC 1999, Vancouver, Canada, vol. 2, pp. 891–895 (June 1999)
44. Oliveira, C., Kim, J.B., Suda, T.: An adaptive bandwidth reservation scheme for high-speed multimedia wireless networks. IEEE Journal on Selected Areas in Communications 16(6), 858–874 (1998)
45. Aljadhai, A., Znati, T.F.: Predictive mobility support for QoS provisioning in mobile wireless networks. IEEE Journal on Selected Areas in Communications 19(10), 1915–1930 (2001)
46. Acampora, A., Naghshineh, M.: An architecture and methodology for mobile-executed handoff in cellular ATM networks. IEEE Journal on Selected Areas in Communications 12(8), 1365–1375 (1994)
47. Ross, S.M.: Stochastic Process, 2nd edn. American Mathematical Society (1997)
48. Press, W.H., Teukolsky, S.A., Vetterling, W.T., Flannery, B.P.: Numerical Recipes in C: The Art of Scientific Computing. Cambridge University Press (1992)
49. Iraqi, Y., Boutaba, R.: When is it worth involving several cells in the call admission control process for multimedia cellular networks? In: Proc. IEEE ICC 2001, Helsinki, Finland, vol. 2, pp. 336–340 (June 2001)

50. Mitchell, K., Sohraby, K.: An analysis of the effects of mobility on bandwidth allocation strategies in multi-class cellular wireless networks. In: Proc. IEEE INFOCOM 2001, Anchorage, USA, vol. 2, pp. 1005–1011 (April 2001)
51. Puterman, M.L.: Markov decision processes: Discrete stochastic dynamic programming. John Wiley & Sons (1994)
52. Haas, Z., Halpern, J.Y., Li, L., Wicker, S.B.: A decision-theoretic approach to resource allocation in wireless multimedia networks. In: Proc. ACM 4th Workshop Discrete Alg. Mobile Comput. Commun., Boston, USA, pp. 86–95 (August 2000)
53. Tijms, H.C.: Stochastic Modeling and Analysis: A Computational Approach. John Wiley & Sons (1989)
54. Saquib, M., Yates, R.: Optimal call admission to a mobile cellular network. In: Proc. IEEE VTC 1995, Chicago, USA, vol. 1, pp. 190–194 (July 1995)
55. Chen, D., Hee, S.B., Trivedi, K.S.: Optimal call admission control policy for wireless communication networks. In: Proc. International Conference on Information, Communication and Signal Processing, ICICS 2001, Singapore (December 2001)
56. Gao, Q., Acampora, A.: Performance comparisons of admission control strategies for future wireless networks. In: Proc. IEEE WCNC 2002, Orlando, USA, vol. 1, pp. 317–321 (March 2002)
57. Kwon, T., Choi, Y., Naghshineh, M.: Optimal distributed call admission control for multimedia services in mobile cellular networks. In: Proc. Mobile Multimedia Ccommunication, MoMuC 1998, Berlin, Germany (October 1998)
58. Xiao, Y., Chen, C.L.P., Wang, Y.: An optimal distributed call admission control for adaptive multimedia in wireless/mobile networks. In: Proc. IEEE MASCOTS 2000, San Francisco, USA, pp. 477–482 (August 2000)
59. Kwon, T., Choi, Y., Naghshineh, M.: Call admission control for adaptive multimedia in wireless/mobile networks. In: Proc. ACM WOWMOM 1998, Dallas, USA, pp. 111–116 (October 1998)
60. Kwon, T., Park, I., Choi, Y., Das, S.: Bandwidth adaptation algorithms with multi-objectives for adaptive multimedia services in wireless/mobile networks. In: Proc. ACM WOWMOM 1999, Seattle, USA, pp. 51–59 (August 1999)
61. Choi, J., Kwon, T., Choi, Y., Naghshineh, M.: Call admission control for multimedia service in mobile cellular networks: A markov decision approach. In: Proc. IEEE ISCC 2000, Antibes, France, pp. 594–599 (July 2000)
62. Yoon, I.-S., Lee, B.G.: A distributed dynamic call admission control that supports mobility of wireless multimedia users. In: Proc. IEEE ICC 1999, Vancouver, Canada, pp. 1442–1446 (June 1999)
63. Kwon, T., Choi, J., Choi, Y., Das, S.: Near optimal bandwidth adaptation algorithm for adaptive multimedia services in wireless/mobile networks. In: Proc. IEEE VTC 1999, Amsterdam, Netherlands, vol. 2, pp. 874–878 (September 1999)
64. Yener, A., Rose, C.: Near optimal call admission policies for cellular networks using genetic algorithms. In: Proc. IEEE Wireless 1994, Calgary, Canada, pp. 398–410 (July 1994)
65. Xiao, Y., Chen, C.L.P., Wang, Y.: A near optimal call admission control with genetic algorithm for multimedia services in wireless/mobile networks. In: Proc. IEEE NAECON 2000, Dayton, USA, pp. 787–792 (October 2000)
66. El-Alfy, E.-S., Yao, Y.-D., Heffes, H.: A learning approach for call admission control with prioritized handoff in mobile multimedia networks. In: Proc. IEEE VTC 2001, Rhodes, Greece, vol. 2, pp. 972–976 (May 2001)

67. Li, B., Li, L., Li, B., Cao, X.-R.: On handoff performance for an integrated voice/data cellular system. Wireless Networks 9(4), 393–402 (2003)
68. Fang, Y.: Thinning schemes for call admission control in wireless networks. IEEE Transactions on Computers 52(5), 686–688 (2003)
69. Pavlidou, F.-N.: Two-dimensional traffic models for cellular mobile systems. IEEE Transactions on Communications 42(2/3/4), 1505–1511 (1994)
70. Wu, H., Li, L., Li, B., Yin, L., Chlamtac, I., Li, B.: On handoff performance for an integrated voice/data cellular system. In: Proc. IEEE PIMRC 2002, Lisboa, Portugal, vol. 5, pp. 2180–2184 (September 2002)
71. Wieselthier, J.E., Ephremides, A.: Fixed- and movable-boundary channel-access schemes for integrated voice/data wireless networks. IEEE Transactions on Communications 43(1), 64–74 (1995)
72. Young, M.C., Haung, Y.-R.: Bandwidth assignment paradigms for broadband integrated voice/data networks. Computer Communications Journal 21(3), 243–253 (1998)
73. Chih-Lin, I., Greenstein, L.J., Gitlin, R.D.: A microcell/macrocell architecture for low and high mobility wireless users. IEEE Journal on Selected Areas in Communications 11(6), 885–891 (1993)
74. Rappaport, S.S., Hu, L.-R.: Microcellular communication systems with hierarchical macrocell overlays: Traffic performance models and analysis. Proc. of the IEEE 82, 1383–1397 (1994)
75. Hu, L.-R., Rappaport, S.S.: Personal communication systems using multiple hierarchical cellular overlays. IEEE Journal on Selected Areas in Communications 13(2), 406–415 (1995)
76. Yeung, K.L., Nanda, S.: Channel management in microcell/macrocell cellular radio systems. IEEE Transactions on Vehicular Technology 45(4), 601–612 (1996)
77. Chang, C., Chang, C.J., Lo, K.-R.: Analysis of a hierarchical cellular system with reneging and dropping for waiting new and handoff calls. IEEE Transactions on Vehicular Technology 48(4), 1080–1091 (1999)
78. Marsan, M.A., Ginella, G., Maglione, R., Meo, M.: Performance analysis of hierarchical cellular networks with generally distributed times and dwell times. IEEE Transactions on Wireless Communications 3(1), 248–257 (2004)
79. Jain, R., Knightly, E.W.: A framework for design and evaluation of admission control algorithms in multi-service mobile networks. In: Proc. IEEE INFOCOM 1999, New York, USA, vol. 3, pp. 1027–1035 (March 1999)
80. Yu, F., Leung, V.C.: Mobility-based predictive call admission control and bandwidth reservation in wireless cellular networks. In: Proc. IEEE INFOCOM 2001, Anchorage, USA, vol. 1, pp. 518–526 (April 2001)
81. Lim, S., Cao, G., Das, C.: An admission control scheme for QoS-sensitive cellular networks. In: Proc. IEEE WCNC 2002, Orlando, USA, vol. 1, pp. 296–300 (March 2002)
82. Soh, W.-S., Kim, H.S.: Qos provisioning in cellular networks based on mobility prediction techniques. IEEE Communications Magazine 41(1), 86–92 (2003)
83. Shen, X., Mark, J.W., Ye, J.: User mobility profile prediction: An adaptive fuzzy inference approach. Wireless Networks 6(5), 363–374 (2000)
84. Evci, C., Fino, B.: Spectrum management, pricing, and efficiency control in broadband wireless communications. Proc. of the IEEE 89, 105–115 (2001)
85. Heikkinen, T.: Congestion based pricing in a dynamic wireless network. In: Proc. IEEE VTC 2001, Rhodes, Greece, vol. 2, pp. 1073–1076 (May 2001)

86. Hou, J., Yang, J., Papavassiliou, S.: Integration of pricing with call admission control for wireless networks. In: Proc. IEEE VTC 2001, Atlantic City, USA, vol. 3, pp. 1344–1348 (October 2001)
87. Haung, Y.-R., Lin, Y.-B., Ho, J.-M.: Performance analysis for voice/data integration on a finite-buffer mobile system. IEEE Transactions on Vehicular Technology 49(2), 367–378 (2000)
88. Ye, J., Shen, X., Mark, J.W.: Call Admission Control in Wideband CDMA Cellular Networks by Using Fuzzy Logic. IEEE Transactions on Mobile Computing 4(2) (March/April 2005)
89. Chen, Y.H., Chang, C.J., Shen, S.: Outage-based fuzzy call admission controller with multi-user detection for WCDMA systems. IEE Proc. Commun. 152(5) (October 2005)
90. Chung-Ju, Chang, L.-C., Kuo, Y.-S., Chen, S.S.: Neural Fuzzy Call Admission and Rate Controller for WCDMA Cellular Systems Providing Multirate Services. In: IWCMC 2006, Vancouver, British Columbia, Canada, July 3-6 (2006)
91. Huang, C.-J., Chuang, Y.-T., Yang, D.-X.: Implementation of call admission control scheme in next generation mobile communication networks using particle swarm optimization and fuzzy logic systems. Expert Systems with Applications 35(3), 1246–1251 (2008)
92. Thilakawardana, S., Tafazolli, R.: Efficient Call Admission Control and SchedulingTechnique for GPRS Using Genetic Algorithms. Mobile Communications Research Group, Centre for Communications Systems Research, CCSR (2004)
93. Wang, S.-L., Hou, Y.-B., Huang, J.-H., Huang, Z.-Q.: Adaptive Call Admission Control Based on Enhanced Genetic Algorithm in Wireless/Mobile Network. In: Proceedings of the 18th IEEE International Conference on Tools with Artificial Intelligence (2006)
94. Wang, S., Cui, Y., Koodli, R., Hou, Y., Huang, Z.: Dynamic Multiple-Threshold Call Admission Control Based on Optimized Genetic Algorithm in Wireless/Mobile Networks. IEICE Transactions on Fundamentals of Electronics, Communications and Computer Sciences E91-A(7), 1597–1608 (2008)
95. Rong, B., Qian, Y., Lu, K., Hu, R.Q., Kadoch, M.: Mobile-Agent-Based Handoff in Wireless Mesh Networks: Architecture and Call Admission Control. IEEE Transactions on Vehicular Technology 58(8), 4565–4575
96. Bernard, W.: 30 Years of Adaptive Neural Networks: Perception, Madalines and Back Propagation. Proceedings of the IEEE 78(9) (September 1990)
97. Chang, C.J., Lin, S.Y., Cheng, R.G., Shiue, Y.R.: PSD-based Neural-net Connection Admission Control. In: Proceedings of IEEE Infocom (April 1997)
98. Diaz-Estrella, A., Jurado, A., Sandoval, F.: New Training Pattern Selection Method for ATM Call Admission Neural Control. Electronic Letters 330 (March 1994)
99. Dueld, N.G., Lewis, J.T., O'Connell, N., Russell, R., Toomey, F.: Entropy of ATM Traffic Streams: A Tool for Estimating QoS Parameters. IEEE Journal on Selected Areas in Communications (August 1995)
100. Elwalid, A., Mitra, D.: Effective Bandwidth of General Markovian Traffic Sources and Admission Control of High Speed Networks. In: Proceedings of IEEE Infocom (June 1994)
101. Guerin, R., Ahmadi, H., Naghshineh, M.: Equivalent Capacity and Its Application to Bandwith Allocation in High-Speed Networks. IEEE Journal on Selected Areas in Communications (September 1991)

102. Haykin, S.: Neural Networks, A Comprehensive Foundation. Macmillan Publishing Company (1994)
103. Hees, H., Lucantoni, D.: A Markov Modulated Characterization of Packetized Voice and Data Traffic and Related Statistical Multiplexer Performance. IEEE Journal on Selected Areas in Communications (September 1986)
104. Hertz, J., Krogh, A., Palmer, R.: Introduction to the Theory of Neural Computation. Addison-Wesley Publishing Company (1991)
105. Hiramatsu, A.: ATM Call Admission Control Using a Neural Network Trained with Virtual Output Buffer Method. In: IEEE International Conference on Neural Networks, vol. 6 (1994)
106. Hiramatsu, A.: Training Techniques for Neural Network Applications in ATM. IEEE Communications Magazine (October 1995)
107. Li, S.Q., Hwang, C.L.: Queue Response to Input Correlation Functions: Discrete Spectral Analysis. IEEE/ACT Transactions on Networking (October 1993)
108. Masters, T.: Practical Neural Network Recipes in C++. Academic Press (1993)
109. Morris, R., Samadi, B.: Neural Network Control of Communications Systems. IEEE Transactions on Neural Networks (1994)
110. Nordstrom, E., Carlstrom, J., Gallmo, O., Asplund, L.: Neural Networks for Adaptive Traffic Control in ATM Networks. IEEE Communications Magazine (October 1995)
111. Ogier, R., Plotkin, N.T., Khan, I.: Neural Network Methods with Traffic Descriptor Compression for Call Admission Control. In: Proceedings of IEEE Infocom (March 1996)
112. Plotkin, N.T., Roche, C.: The Entropy of Cells Streams as a Traffic Descriptor in ATM Networks. IFIP Performance of Communications Systems (October 1995)
113. Sarajedini, A., Chau, P.M.: Quality of Service Prediction Using Neural Networks. In: Proceedings of MILCOM, vol. 2 (1996)
114. Tham, C.-K., Soh, W.-S.: Multi-service Connection Admission Control using Modular Neural Networks. In: Proceedings of IEEE Infocom (March 1998)
115. Youssef, S., Habib, I., Saadawi, T.: A Neurocomputing Controller for Bandwdith Allocation in ATM Networks. IEEE Journal on Selected Areas in Communications (February 1997)

Chapter 2
An Overview of Computational Intelligence Algorithms

This chapter provides an overview of selected computational intelligence algorithms, which will be required to understand the rest of the book. It begins with a review of fuzzy sets and logic, and would gradually explore swarm and evolutionary algorithms, and neural nets. The coverage on swarm and evolutionary algorithms include Genetic Algorithm, Particle Swarm Optimization Bio-geography Based Optimization and Differential Evolution algorithm. Supervised, unsupervised and reinforcement learning algorithms will be outlined under neural nets. The chapter ends with scope of applications of computational intelligence algorithms in call admission.

2.1 Introduction

The chapter introduces the foundations of computational intelligence techniques in a nutshell. Computational intelligence is a vast family of knowledge comprising of several models and techniques, including the logic of fuzzy sets, neurocomputing, and swarm and evolutionary computation. The logic of fuzzy sets provides a frame work for uncertainty management in complex reasoning systems. Neural nets offer automatic techniques of machine learning and pattern recognition. Evolutionary algorithms are widely being used for intelligent search, optimization and machine learning. Swarm algorithms are nature-inspired techniques capable of imitating nature by judiciously selecting their power of natural optimization in identifying food sources. Researchers are taking keen interest to develop newer bio-inspired optimization algorithms by imitating the behavior of lower level creatures such as ants, bees and swaps.

The chapter begins with logic of fuzzy sets. It introduces membership functions and fuzzy relations with special reference to implication relations. It also reviews fuzzy reasoning. Principles of machine learning are introduced through artificial neural networks. The chapter provides the basis of learning in biological nervous system, and its electrical equivalent model, that too stems from its biological counterpart. It also briefly overviews supervised and unsupervised learning, the basic two learning policies that humans normally employ in their natural learning process. The later part of the chapter overviews a few swarm and evolutionary algorithms, covering Particle Swarm Optimization (PSO), Biogeography Based Optimization (BBO) and Genetic Algorithm and Differential Evolution algorithm (DE).

S. Ghosh and A. Konar: CAC in a Mobile Cellular Network, SCI 437, pp. 63–94.
springerlink.com   © Springer-Verlag Berlin Heidelberg 2013

2.2   A Review of Fuzzy Sets and Logic

In a conventional set, the condition defining the set boundaries is very rigid. For example, consider a universal set AGE, OLD, VERY OLD, YOUNG, CHILD and BABY are subsets of the universal set AGE. The conventional approach to define these sets is illustrated below:

$$BABY= \{age \in AGE: 0 \ year \leq age <1year\},$$

$$CHILD= \{age \in AGE: 1 \ year \leq age \leq 10years\},$$

$$YOUNG= \{age \in AGE: 19 \ years \leq age \leq 40years\},$$

$$OLD= \{age \in AGE: 60 \ years \leq age < 80years\},$$

$$and \ \ VERY \ OLD= \{age \in AGE: 80 \ years \leq age < 120 \ years\}.$$

In the above definitions age is a variable that may presume any value in the range [0, 120] years. It is clear from the definition that the boundary of each set is distinct. Thus an age=11 months 29 days is a member of the set BABY, but once it is 1 year it falls in the set CHILD. Thus there is a sharp demarcation in the boundary definition of the sets BABY and CHILD at age=1 year. Measurements in a real world system being highly imprecise, such a sharp demarcation of 2 set boundaries may cause a wrong allocation of the members to a given set.

Another characteristic of a conventional set includes assignment of a grade of membership 1 to all its members and 0 to all its non-members. The following connotation is used to describe that the membership of an element x in a set A is 1, and the membership of a non-element y in the set A is 0.

$$\mu_A(x) = 1 \tag{2.1}$$

$$\mu_A(y) = 1 \tag{2.2}$$

A fuzzy set extends the binary membership: {0,1} of a conventional set to a spectrum in the interval of [0, 1]. Further, unlike a conventional set, all elements of the universal set U are members of a given set A. Thus for each element $x \in U$,

$$0 \leq \mu_A (x) \leq 1 . \tag{2.3}$$

It needs mention here that as all elements of a universal set U are members of a given fuzzy set A, therefore, 2 fuzzy sets A and B may have an overlap in the boundary definitions. For example, in contrast to the respective conventional sets: BABY, CHILD, YOUNG, OLD and VERY OLD, the corresponding fuzzy sets allow any age in the interval [0, 120] years as a member of each of the above sets but with different memberships in [0, 1]. As a specific instance, the age 22 is a member of all the fuzzy sets but the membership of age (=22) to belong to the sets BABY, CHILD, YOUNG, OLD and VERY OLD respectively are 0.001, 0.01,1.00, 0.60 and 0.20. The above example makes sense in the line of reasoning that an age of 22 corresponds to a young person, so the membership of age (=22) to be young is high (1.00). The relative grading of the other memberships thus can be easily understood from the usual meaning of the terms BABY, CHILD, OLD and VERY OLD. A fuzzy set thus can be formally defined as follows.

Definition 2.1: A fuzzy set A is a set of ordered pairs, give

$$A = \{x, \ \mu_A(x) : x \in X\} \tag{2.4}$$

where X is a universal set of objects (also called the universe of discourse) and $\mu_A(x)$ is the grade of membership of the object x in A. Usually, $\mu_A(x)$ lies in the closed interval of [0,1].

It may be added here that some authors [1] relax the range of membership from [0, 1] to [0, Rmax] where Rmax is a positive finite real number. One can easily convert [0, Rmax] to [0, 1] by dividing the membership values in the range [0, Rmax] by Rmax.

There are other notations of fuzzy sets as well. For instance, the ordered pair *(x, $\mu_A(x)$)* in the definition of fuzzy set is also written as $x/\mu_A(x)$ or μ_A/A as well. Let the elements of set X be $x_1, x_2, \ldots \ldots x_n$. Then the fuzzy set $A \subseteq X$ is denoted by any of the following nomenclature.

$$A = \{(x_1, \mu_A(x_1)), (x_2, \mu_A(x_2)), \ldots \ldots (x_n, \mu_A(x_n))\}$$

Or $\quad A = \{x_1/\mu_A(x_1), x_2/\mu_A(x_2), \ldots \ldots x_n/\mu_A(x_n)\}$

Or $\quad A = \{x_1/\mu_A(x_1) + x_2/\mu_A(x_2) + \ldots \ldots + x_n/\mu_A(x_n)\}$

Or $\quad A = \{\mu_A(x_1)/x_1 + \mu_A(x_2)/x_2 + \ldots \ldots + \mu_A(x_n)/x_n\}$

Or $\quad A = \{\mu_A(x_1)/x_1, \mu_A(x_2)/x_2, \ldots \ldots \mu_A(x_n)/x_n\}$

In this book we used the last option. The details of membership function $\mu_A(x)$ is formalized below.

2.2.1 Membership Functions

The grade of membership $\mu_A(x)$ maps the object or its attribute x to positive real numbers in the interval [0, 1]. Because of its mapping characteristics like a function, it is called membership function. A formal definition of the membership function is given below for the convenience of the readers.

Definition 2.2: A membership function $\mu_A(x)$ is characterized by the following mapping:

$$\mu_A : x \to [0,1], \qquad x \in X \tag{2.5}$$

where x is a real number describing an object or its attribute and X is the universe of discourse and A is a subset of X.

A question that naturally arises is: how to construct a membership function? The following examples provide a thorough insight to the selection of the membership functions.

Example 2.1: Consider the problem of defining BABY, CHILD, YOUNG, OLD and VERY OLD by membership functions. The closer the age of a person to 0, the higher is his/her membership to be a BABY. So, if x is the age of the person, we can define BABY as follows:

$$BABY = \{x, \mu_{BABY}(x)\} \quad where \quad \mu_{BABY}(x) = \exp(-x) \quad (2.6)$$

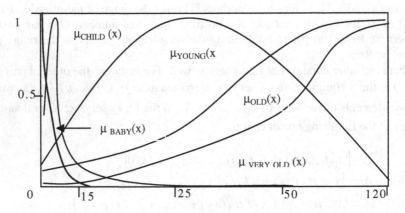

Fig. 2.1 Membership curves for the fuzzy sets: BABY, CHILD, YOUNG, OLD and VERY OLD. The x-axis denotes the age in years and the y-axis denotes the memberships of the given fuzzy sets at different ages.

Thus as $x \rightarrow 0$, $\mu_{BABY}(x) \rightarrow 1$. Further, as x increases, $\mu_{BABY}(x)$ decreases exponentially. The membership function $\mu_{BABY}(x)$ can also be designed to have a controlled decrease with increasing x by including a factor α to x in exp(-x). Thus,

$$\mu_{BABY}(x) = \exp(-\alpha x) \quad for \quad \alpha > 0 \quad (2.7)$$

Larger the value of α, the higher is the falling rate of $\mu_{BABY}(x)$ over x. In a similar manner we can define the membership functions for CHILD, YOUNG, OLD and VERY OLD fuzzy sets. But before representing them mathematically let us take a look at them.

The membership curves for the fuzzy sets: BABY, CHILD, YOUNG, OLD and VERY OLD are shown in Fig. 2.1. The curve for CHILD fuzzy set has the peak at some age slightly greater than 0 and has a sharp fall off around the peak. The logical interpretation of this directly follows from the meaning of the word child.

The membership curve for the fuzzy set YOUNG has a peak at age x=25 and falls off very slowly on both sides around the peak.

As youth is the most charming period of the human beings, we prefer to call people YOUNG even if they are away from 25 on either side. If the readers' view is different they can allow a sharp falloff of the curve around the age x=25. One interesting point to note about the OLD and VERY OLD membership curves is that OLD curve throughout has a higher membership than the VERY OLD curve until both saturate at age x = 120 years. This is meaningful because if someone is called VERY OLD then he must be OLD, but the converse may not be true.

There are many ways to represent the membership functions shown in Fig. 2.1 by mathematical functions. One such representation is given below:

$$\mu_{CHILD}(x) = ax^2/(1 + bx^2 + xc), \qquad a, b, c > 0 \qquad (2.8)$$

$$\mu_{YOUNG}(x) = exp[-(x-25)^2/2\sigma^2], \qquad \sigma > 0 \qquad (2.9)$$

$$\mu_{OLD}(x) = 1 - exp(dx^2) \qquad d > 0 \qquad (2.10)$$

$$and \ \mu_{VERYOLD}(x) = 1 - exp(-dx), \qquad d > 0 \qquad (2.11)$$

The parameters a, b, c and d in the above membership functions are selected intuitively by the experts based on their subjective judgement in the respective domains. Tuning of these parameters is needed to control the curvature and sharp changes on the curves around some selected x-values.

2.2.2   Continuous and Discrete Membership Functions

The universe of discourse (or simply the universe) of a fuzzy set may exist for both discrete and continuous spectrum. For example, the roll number of students in a class is a discrete universe. On the other hand the height of persons is a continuous universe as height may take up any values between 4' to 8'. It may be mentioned here that a continuous universe is sometimes sampled at regular or irregular intervals for using it as a discrete universe. The membership curve of YOUNG in Fig. 2.1 may be, for instance, discretized at age x= 18, 22, 24, 28. This is an example of non-uniform/ irregular sampling as the intervals of sampling 18-22, 22-24, 24-28 are unequal. The membership curve of YOUNG may alternatively be sampled at a regular interval of age x=18, 20, 22, 24, say. This is an example of uniform/ regular sampling. Fig. 2.2(a) and (b) describe the instances of the non-uniform and uniform sampling of the YOUNG membership curve.

2.2.3   Fuzzy Implication Rules

Reasoning informally refers to generation of inferences from a given set of facts and rules. The subject logic is concerned with the formalization of methodology and principles of reasoning. Production rules are synonymously called implication rules in traditional logic. Implication rules of traditional logic are extended with fuzzy linguistic variables. In this section, the implication rules in the classical (propositional/ predicate) logic and their extension in fuzzy domain are introduced.

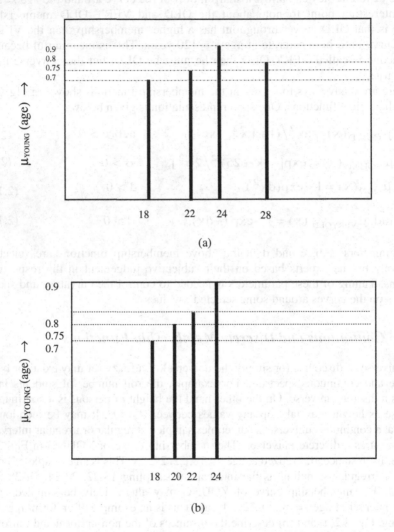

Fig. 2.2 (a) Non-uniform and (b) uniform sampling of the YOUNG membership curve.

Let us try to formalize the rule: IF x is a banana and x is yellow THEN x is ripe in predicate logic. For this formalization, we define 3 predicates Banana(x), Yellow(x) and Ripe(x), where each of these predicates has only 2 possible truth values: true or false. Using a IF-THEN (implication) operator, the above rule can be written as:

$$Banana(x), Yellow(x) \rightarrow Ripe(x).$$

where the comma in the left side of the implication sign ($\rightarrow$) denotes logical conjunction (AND) of the antecedent predicates.

In order to allow instantiation of the antecedent predicates with the contents of the WM, we consider 6 fuzzy sets: YELLOW, VERY-YELLOW, MORE-OR-LESS-YELLOW, RIPE, VERY-RIPE and MORE-OR-LESS-RIPE. Fuzzy extensions of the last rule then may be re-stated as follows:

> *Rule 1: IF a banana is YELLOW*
> *THEN it is RIPE.*
> *Rule 2: IF a banana is VERY-YELLOW*
> *THEN it is VERY-RIPE.*
> *Rule 3: IF a banana is MORE-OR-LESS-YELLOW*
> *THEN it is MORE-OR-LESS-RIPE.*

It may be added here that unlike production systems, the logic of fuzzy sets allows firing of these 3 rules concurrently in presence of a data element concerning color of a banana in the WM. Thus conflict resolution is not employed in fuzzy logic.

Formally, let x be a linguistic variable in a universe X, and A1, A2 and A3 are three fuzzy sets under the universe X. Also assume that y be another linguistic variable in a universe Y and B1 , B2 and B3 are 3 fuzzy sets under Y. Then the implication rules between variable x and y may be described as

> *Rule1: IF x is A1 THEN y is B1.*
> *Rule2: IF x is A2 THEN y is B2.*
> *Rule 3: IF x is A3 THEN y is B3.*

Suppose the WM contains x is A', where A' is semantically close to A1, A2 and A3 respectively. In the subsequent sections, we demonstrate the evaluation procedure of y is B' from the known membership distributions of x is A_1, y is B_1, x is A_2, y is B_2, x is A_3, y is B_3 and x is A'

2.2.4 Fuzzy Implication Relations

A fuzzy implication relation [2] for a given rule: IF x is Ai THEN y is Bi is formally denoted by

$$R_i(x, y) = \left\{ \mu_{R_i}(x, y) / (x, y) \right\} \tag{2.12}$$

where the membership function $\mu_{R_i}(x, y)$ is constructed intuitively by many alternative ways. One typical implication relations is presented below.

Mamdani Implication: Mamdani proposed the following two implication functions [7, 12].

$$\mu_{R_i}(x, y) = Min[\mu_{A_i}(x), \mu_{B_i}(y)] \tag{2.13}$$

$$OR \ \mu_{R_i}(x, y) = \mu_{A_i}(x\mu_{A_i}(x)\mu_{B_i}(y) \tag{2.14}$$

Mamdani implication functions are most widely used implications in fuzzy systems and fuzzy control engineering. This implication relation is constructed based on the assumption that fuzzy IF-THEN rules are local. For example, consider the implication rule: IF height is TALL THEN speed is HIGH. By Mamdani's implication function, we do not want to mean: IF height is SHORT THEN speed is SLOW. The second rule is rather an example of a non-local rule. The knowledge engineer thus has to decide whether he prefers local or non-local rules. If he prefers local rules then Mamdani's implication relation should be used.

2.2.5 Fuzzy Logic

The logic of fuzzy sets, also called fuzzy logic, is an extension of the classical propositional logic from 2 perspectives. First, instead of binary valuation space (truth/falsehood) of the propositional logic, fuzzy logic provides a multi-valued truth-space in [0, 1]. Secondly, propositional logic generates inferences based on the complete matching of the antecedent clauses with the available data, whereas fuzzy logic is capable of generating inferences even when a partial matching of the antecedent clauses against data elements in WM exists. In this section, we present 3 typical propositional inference rules and describe their possible extensions in fuzzy logic.

2.2.6 Typical Propositional Inference Rules

Let p, q and r be 3 propositions. The following 3 propositional inference rules are commonly used for logical inferencing.

➢ Modus Ponens: Given a proposition p and a propositional implication rule p→q, we can derive the inference q. Symbolically,

$$p \wedge (p \to q) \Rightarrow q \tag{2.15}$$

where ⇒ denotes a logical provability operator. This above inference rule is well known as modus ponens.

➢ Modus Tollens: Given a proposition ¬q, and the implication rule p→q, we can derive the inference ¬p. symbolically,

$$\neg q \wedge (p \to q) \Rightarrow \neg p \tag{2.16}$$

The above inference rule is known as modus tollens.

> ➤ Hypothetical Syllogism: Given 2 implication rules p→q and q→r, then we can easily derive a implication p→r. Symbolically,

$$(p \to q) \land (q \to r) \Rightarrow p \to r \qquad (2.17)$$

The above inference rule is popularly known as hypothetical syllogism or chain rule.

2.2.7 Fuzzy Extension of the Inference Rules

The logic of fuzzy set provides a general framework for the extension of the above 3 propositional inference rules. Fuzzy extension of modus ponens, modus tollens and hypothetical syllogisms are called generalized modus ponens, generalized modus tollens and generalized hypothetical syllogism respectively.

Generalized Modus Ponens (GMP): Consider a fuzzy production rule: IF x is A then y is B, and a fuzzy fact: x is A'. The GMP inference rule then infers y is B'. Here A, B, A' and B' are fuzzy sets such that A' is close to A, and B' is close to B. The inference rule also states that the closer the $A/$ to A, the closer the B' to B. Symbolically, the GMP can be stated as follows:

> *Given:* *IF x is A THEN y is B.*
> *Given:* *x is A'*
> _____
> *Inferred:* *y is B'*

Generalized Modus Tollens (GMT): Given a fuzzy production rule: IF x is A THEN y is B, and a fuzzy fact y is B', the GMT then infers x is A', where the more is the difference between B' and B, the more is the difference between A' and A. Symbolically, the GMT is stated as follows:

> *Given:* *IF x is A THEN y is B.*
> *Given:* *y is B'*
> _____
> *Inferred:* *x is A'.*

Generalized Hypothetical Syllogism (GHS): Given 2 fuzzy production rules: IF x is A THEN y is B, and IF y is B' THEN z is C', where A, B and C are 3 fuzzy sets, and B' is close to B. Then the GHS infers : If x is in A then z is in C'. The closer the B' to B, the closer the C' to C. Symbolically, we can state this rule as follows:

> *Given: IF x is A THEN y is B.*
> *Given: IF y is B' THEN z is C*
> _____
> *Inferred: If x is in A then z is in C'.*

In the above definition of the inference rules, we just mentioned that A' is close to A, B' is close to B and the like. But we did not mention what we exactly mean by "close to". In fact "close to" can take any of the following forms: VERY, VERY-VERY, MORE-OR-LESS, NOT, ABOUT-TO, AROUND and other fuzzy hedges that mean a fuzzy set A/ are approximately similar to A.

2.2.8 The Compositional Rule of Inference

In this section we present the methodology for the evaluation of fuzzy inferences for GMP, GMT and GHS. This, however, calls for formalization of a fuzzy rule called the compositional rule of inference. The compositional rule of inference is usually applied to 2 fuzzy membership distribution, one of which usually have a smaller number of linguistic variables. The rule extends the latter membership distribution cylindrically, so as to increase its number of linguistic variables to the former distribution. The intersection of the former and the resulting distribution is then projected to desired axes. The whole process is referred to as the compositional rule of inference. How exactly the compositional rule of inference is applied to determine the fuzzy inferences in GMP, GMT and GHS is presented below.

2.2.9 Computing Fuzzy Inferences Using GMP

Given a fuzzy production rule: IF x is A THEN y is B, and a fact y is B', we by GMT infer x is A'. In this section we present the principle of determining the membership distribution of x is A/, $\mu_A/(x)$, from the membership distribution of y is B', $\mu_B'(y)$, and the membership $\mu_R(x, y)$ of the fuzzy relation between the antecedent and consequent part of the given rule. The computation involves cylindrical extension of $\mu_B'(y)$ to μ_B' CYL (x, y), then intersection of μ_B' CYL (x, y) with $\mu_R(x, y)$ and finally projection of the resulting relation ox X-axis. Thus, following the same steps as in the case of GMP, it can easily be shown that

$$\mu_{A'}(x) = \max_{y \in Y}\left[t\{\mu_{B'}(y), \mu_R(x, y)\}\right]. \qquad (2.18)$$

When $\mu_B'(y)$ and $\mu_R(x, y)$ are discrete relations represented by a row vector and a matrix respectively, we can represent the above result by the following max-min composition operation:

$$\mu_{A'}(x) = \mu_{B'}(y) \circ \left[\mu_R(x, y)\right]^T \qquad (2.19)$$

where T denotes the transposition operator over the given relation.

Example 2.2: This example illustrates the computation of GMP using max-min composition operator.

We take the same $\mu_R(x, y)$ as in Example 2.1, and the $\mu_B/(y)$ we obtained as the result in that example, and plan to determine $\mu_A/(x)$ by the compositional rule of inference. Thus,

$$\mu_{A'}(x) = \mu_{B'}(y) \circ [\mu_R(x, y)]^T = [0.8 \ 0.6 \ 0.9] \circ \begin{bmatrix} 0.8 & 0.6 & 0.5 \\ 0.6 & 0.5 & 0.9 \\ 0.7 & 0.6 & 0.5 \end{bmatrix}^T$$

$$= [0.8 \ 0.9 \ 0.7]$$

It may be noted that $\mu_{A'}(x)$ thus obtained is not same as that presumed in Example 2.2.

2.3   Neural Nets

This section provides an introduction to biological neural networks and their mathematical models called artificial neural networks. Neural networks have proved them successful in pattern recognition, system identification, function approximation and many others.

2.3.1   Biological Neural Networks

The human nervous system consists of small unicellular structures called neurons. Neurons thus are fundamental units or building blocks of a biological nervous system. A neuron comprises of 5 elements: dendrite, Dendron, cell body, synapse and axon (Fig. 2.3). The dendrites receive signals from other neurons, muscles or sensory organs and carry them to a relatively thick fiber called Dendron. The Dendron carry the signals to the cell body, which in turn generates a composite signal based on the strength of the signals received from the dendrites. The composite signal is transmitted to the synapse through the axon (Fig. 2.3).

The synapse is like a potential barrier that controls the flow of signal from the axon of one neuron to the dendrites of other neurons. The transmission of signals from one neuron to another at a synapse is a complex chemical process. Generally, a specific chemical called neuro-transmitter is released from the transmitting end of the synapse. The transmitters lower or raise the potential barrier of the synapse. On lowering the potential barrier, the signal from the transmitting end can easily reach the other end of the synapse. On raising the barrier, a signal loss takes place and only a small component of the signal can reach the other end of the synapse. Flow of signal from one neuron to the others thus can be controlled by the synapse. When the influence of the synapse tends to activate the post-synaptic neuron, the synapse is called excitatory. On the other hand, when the synapse prohibits the passage of signal flow to the post-synaptic neuron, the synapse is called inhibitory. The synaptic ending of an axon thus is of excitatory or inhibitory nature.

How the neurons participate in the learning process of the human beings remained a mystery till this date. However, it is evident that the process of learning has a correspondence with the type and amount of neuro-transmitters released at the pre-synaptic end of the neurons. Thus for similar sensory/control/

motor actions a neuron releases the same type and amount of neuro-transmitters. How exactly the neurons perform the above task is an interesting topic of research for the biologists and medical researchers.

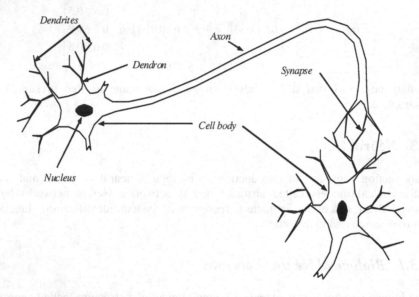

Fig. 2.3 A Biological Neural Net comprising of two neurons, where the dendrites of the second neuron receive signal from the synapse of the first neuron.

2.3.2 Artificial Neural Networks

Artificial neural networks are electrical analogue of the biological nervous system. A typical artificial neuron is mathematically represented by two modules: i) a linear activation/ inhibition module and ii) a non-linearity that limits the signal levels within a finite band. Fig. 2.4 presents the typical organization of an artificial neuron.

The summer in Fig. 2.4 takes the role of the cell body and the inputs of the summer may be treated like dendrites. The synapse is modeled by a non-linear function and the connection from the summer to the non-linear unit is like the axon. Here, Net is a linear combiner of the inputs $x_1, x_2, ..., x_n$. Mathematically,

$$Net = \sum_{i=1}^{n} w_i x_i \tag{2.20}$$

where some of the inputs are excitatory (positive) and the rest are inhibitory (negative). 'Out' in the present context can take different mathematical forms. Some of the common forms are presented below.

$$Out = u(Net - th) \tag{2.21}$$

$$Out = Sgn(Net) \tag{2.22}$$

$$Out = 1/[1 + \exp(-Net)] \tag{2.23}$$

$$Out = \tanh(Net/2) \tag{2.24}$$

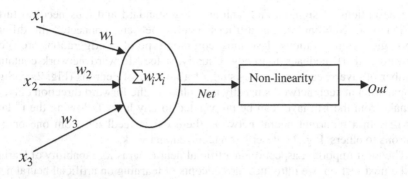

Fig. 2.4 A Typical Artificial Neuron

The mathematical form of Out can be smooth functions like sigmoid vide expression (2.23) or tanh vide expression (2.24) and sharp changing functions like step vide expression (2.21) or signum function (vide expression (2.22))

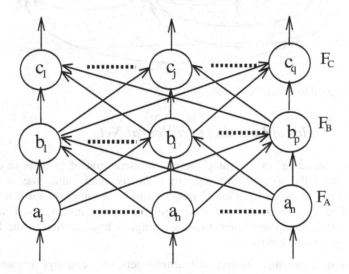

Fig. 2.5 A Feed-Forward Neural Net of 3 Layers

The unit step function u in expression (2.21) is formally defined as follows:

$$u(net - th) = \begin{cases} 1, & Net > th \\ 0, & otherwise \end{cases} \tag{2.25}$$

The signum function Sgn in expression (2.22) is formally defined as

$$Sgn(Net) = \begin{cases} +1, & Net > 0 \\ -1, & otherwise \end{cases} \qquad (2.26)$$

The definitions of sigmoid and tanh are very standard and thus need no further explanation. Neurons in an artificial neural net are connected in different topological configurations. Two most common type of configurations are i)feed-forward and ii) feedback topology. Usually, a feed-forward network contains a number of layers, each layer consisting of a number of neurons (Fig. 2.4). Signal propagation in such networks usually take place in the forward direction only, i.e., signals from the i-th layer can be propagated to any layer following the ith layer, for i≥1. In a recurrent neural network, there exists feedback from one or more neurons to others. Fig. 2.6 describes a recurrent network.

The most important aspect of an artificial neural net is its capability of learning. In the next section, we introduce the concepts of learning on artificial neural nets.

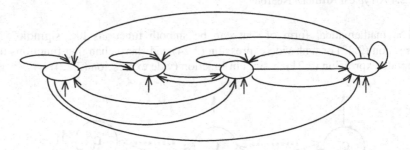

Fig. 2.6 A Typical Recurrent Neural Net

2.3.3 *Principles of Learning in a Neural Net*

Informally "encoding" or "learning" refers to adaptation of weights in a neural net. Thus until the weights converge to a steady state value, the process of encoding is continued. Adaptation of weights can be accomplished in a neural net by 4 different ways; they are supervised learning, unsupervised learning, reinforcement learning and competitive learning. A brief outline to the learning schemes is presented below.

♦ **Supervised Learning:** Supervised learning generally employs a trainer, who provides the input-output training instances of a given neural net. As an example, let us consider a pattern recognition problem, where we need to recognize an object from its feature- space. Here, the set of features such as size of the object and its shape described by its boundary descriptors, for instance, may be considered as input, while the type of the object such as books, pencils etc. may be treated as output of the neural net. Thus for n distinct objects, we require n-outputs of the neural net, each corresponding to one object.

When one of n-outputs has the maximum value, the object is regarded to fall within the particular class. Further for a large n, we can denote the output class by an encoded number, such as binary string. Thus for a given input feature vector, if a binary string 0011 appears at the output, we consider the object to belong to class 3.

Fig. 2.7 describes a scheme for supervised learning. Here, given a input vector I and a target vector T, we need to fix the weights in the network, such that T is produced at the output of the network when excited with the input I. How can we achieve this? First, we initialize the weights randomly. Then for the given input vector I, suppose the network generates the vector O at its output. An error vector E = T – O is then generated, and a supervised learning algorithm is used to adjust the Network parameters based on the error vector.

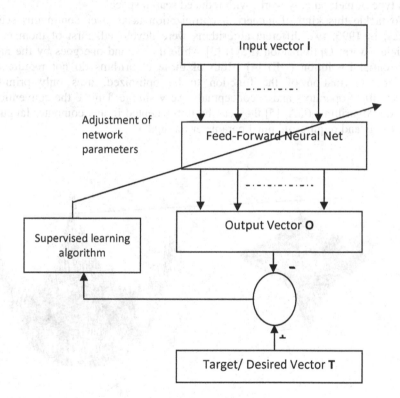

Fig. 2.7 A Simple Supervised Learning Scheme.

2.4 Swarm and Evolutionary Algorithms

Problems which involve global optimization over continuous spaces are ubiquitous throughout the scientific community. In general, the task is to optimize certain properties of a system by pertinently choosing the system parameters. For convenience, a system's parameters are usually represented as a vector.

The standard approach to an optimization problem begins by designing an objective function that can model the problem's objectives while incorporating any constraints. In a complex real life search problem, the search space may be a rough landscape, riddled with multiple local maxima/minima.

The objective function is very often non differentiable and/or discontinuous at a number of points. As for example consider the following functions shown in Fig. 2.8. Since the derivative based methods are of no help, other methods, combining mathematical analysis and random search came up for them. Imagine you scatter small robots in a Mountainous landscape. Those robots can follow the steepest path they found. When a robot reaches a peak, it claims that it has found the optimum. This method of hill climbing is very efficient, but there's no proof that the optimum has been found, each robot can be blocked in a local optimum. This type of method only works with reduced search spaces.

To tackle this kind of numerical optimization tasks over continuous search spaces, in 1995, two different algorithms were developed. First of them is the Particle Swarm Optimization (PSO) [3] while the second one goes by the name Differential Evolution (DE) [4]. Both of these algorithms do not require any gradient information of the function to be optimized, uses only primitive mathematical operators and is conceptually very simple. Unlike the conventional Genetic Algorithms (GA) [5] they can be implemented in any computer language very easily and requires minimal parameter tuning.

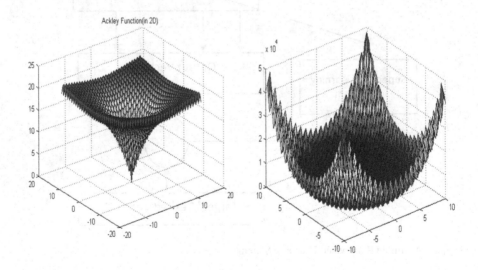

Fig. 2.8 Functions with Huge Number of Local Minima and Maxima

Their performance does not deteriorate severely with the growth of the search space dimensions as well. These issues perhaps have a great role in the popularity of the algorithms within the domain of machine intelligence and cybernetics.

2.4.1 Classical PSO

The concept of function-optimization by means of a particle swarm was introduced by James Kennedy and Russel C. Eberhart in an IEEE neural network conference paper from 1995 [3]. Suppose the global optimum of an n-dimensional function is to be located. The function may be mathematically represented as $f(x_1, x_2, x_3, ..., x_n) = f(\vec{X})$ where $\vec{X}$ is called the parameter vector which actually represents the set of independent variables. The task is to find out such a $\vec{X}$, that the function value $f(\vec{X})$ is either a minimum or a maximum denoted by f* in the search range. If the components of $\vec{X}$ assume real values then the task is to locate a particular point in the n dimensional hyperspace which is a continuum of such points.

Example 2.3 Consider the simplest two dimensional sphere function given by,

$$f(x_1, x_2) = f(\vec{X}) = x_1^2 + x_2^2$$

If x_1 and x_2 can assume real values only then by inspection it is pretty clear that the global minima of this function is at $x_1=0$, $x_2=0$ i.e. at the origin (0, 0) of the search space and the minimum value is f(0, 0) = f* = 0. No other point can be found in the x_1-x_2 plane at which value of the function is lower than f* = 0. Now the case of finding the optima is not so easy for functions like this one:

$$f(x_1, x_2) = x_1 \sin(4\pi x_2) - x_2 \sin(4\pi x_1 + \pi) + 1 ;$$

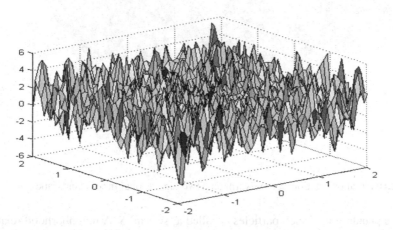

Fig. 2.9 Surface plot of the above-mentioned function

This function has multiple peaks and valleys and a rough fitness landscape. A surface plot of the function is shown in Fig. 2.9. To locate the global optima quickly on such a rough surface calls for parallel search techniques. Here many agents start from different initial locations and go on exploring the search space until some (if not all) of the agents reach the global optimal position. The agents may communicate among themselves and share the fitness function values found by them.

PSO is in principle such a multi-agent parallel search technique. Particles are conceptual entities which fly through the multi-dimensional search space. At any particular instant each particle has a position and a velocity. The position vector of a particle with respect to the origin of the search space represents a trial solution of the each problem.

At the beginning a population of particles is initialized with random positions marked by vectors $\vec{X}_i$ and random velocities $\vec{V}_i$. Initial distribution of particles on a two dimensional search space may be illustrated in Fig. 2.10.

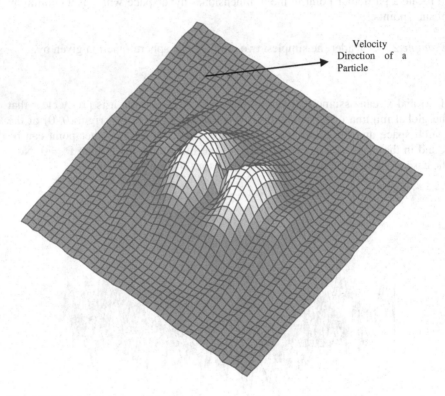

Velocity Direction of a Particle

Fig. 2.10 Initial orientation of the swarm on a two dimensional fitness landscape

The population of such particles is called a 'swarm' S. A neighborhood relation N is defined in the swarm. N determines for any two particles P_i and P_j whether they are neighbors or not. Thus for any particle P, a neighborhood can be assigned as $N(P)$, containing all the neighbors of that particle. Each particle P has two state variables:

> Its current position $\vec{x}(t)$
> Its current velocity $\vec{v}(t)$.

And also a small memory comprising,

> Its previous best position $\vec{p}(t)$ i.e. personal best experience.
> The best $\vec{p}(t)$ of all $P \in N(P)$: $\vec{g}(t)$ i.e. the best position found so far in the neighborhood of the particle.

The best $\vec{p}(t)$ of all $P \in N(P)$: $\vec{g}(t)$ i.e. the best position found so far in the neighborhood of the particle. The PSO scheme has the following algorithmic parameters:

> V_{max} or maximum velocity which restricts $\vec{V}_i(t)$ within the interval $[-v_{max}, v_{mas}]$.
> An inertial weight factor ω [6].
> Two uniformly distributed random numbers φ_1 and φ_2 which respectively determine the influence of $\vec{p}(t)$ and $\vec{g}(t)$ on the velocity update formula.
> Two constant multiplier terms C_1 and C_2 known as "self confidence" and "swarm confidence" respectively.

Initially the settings for $\vec{p}(t)$ and $\vec{g}(t)$ are $\vec{p}(0) = \vec{g}(0) = \vec{x}(0)$ for all particles. Once the particles are initialized, the iterative optimization process begins where the positions and velocities of all the particles are altered by the following recursive equations. The equations are presented for the dth dimension of the position and velocity of the i-th particle.

$$V_{id}(t+1) = \omega V_{id}(t) + C_1 \varphi_1 (P_d(t) - X_{id}(t)) + C_2 \varphi_2 (g_d(t) - X_{id}(t)) \qquad (2.27)$$
$$X_{id}(t+1) = X_{id}(t) + V_{id}(t+1)$$

The first term in the velocity updating formula represents the inertial velocity of the particle. The second term $\vec{P}(t)$ involving represents the personal experience of each particle and is referred to as "cognitive part".

The last term of the same relation is interpreted as the "social term" which represents how an individual particle is influenced by the other members of its society. The velocity updating scheme has been presented in Fig. 2.11, using a humanoid agent in place of a particle on the spherical functional surface. After having calculated the velocities and position for the next time step $t+1$, the first iteration of the algorithm is completed.

Typically, this process is iterated for a certain number of time steps, or until some acceptable solution has been found by the algorithm or until an upper limit of CPU usage has been reached. Once the iterations are terminated, most of the particles are expected to converge to a small radius surrounding the global optima of the search space. The ideal distribution of the particles after the algorithm is stopped has been shown in Fig. 2.12.

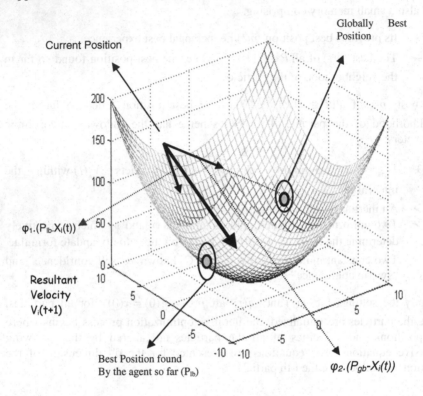

Fig. 2.11 Illustrating the velocity updating scheme of basic PSO

The algorithm can be summarized in the following pseudo code:

Procedure Particle_swarm_optimization
set t = 0;
Initialize φ_1, φ_2, V_{max} and define N;

***While** (termination_condition = FALSE)*
{
    $\forall p \in S$: calculate $\vec{v}(t+1)$ and $\vec{x}(t+1)$ using equations (1); $\forall p \in S$: update
$\vec{p}(t+1)$ with $\vec{x}(t+1)$ if $f(\vec{x}(t+1))$ is better than $f(\vec{x}(t))$
    $\forall p \in S$: update $\vec{g}(t+1)$ with the best $\vec{p}(t+1)$ in N(p);
}

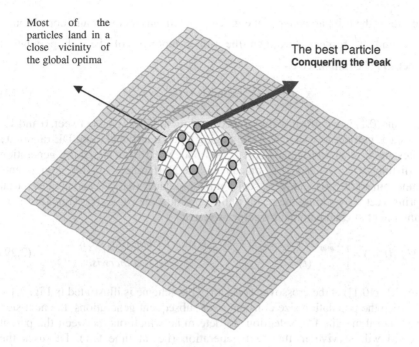

Most of the particles land in a close vicinity of the global optima

The best Particle
Conquering the Peak

Fig. 2.12 Ideal distribution of the particles on a two dimensional fitness landscape after the algorithm is terminated

2.4.2 Differential Evolution

In 1995 Storn and Price made an attempt to replace the classical crossover and mutation operators in GA by alternative operators [4], and found a suitable vector differential operator to handle the problem. They proposed a new algorithm based on this operator, and called it Differential Evolution (DE). DE searches for a global optimum in a D-dimensional hyperspace. It begins with a randomly initialized population of D-dimensional real-valued parameter vectors. Each vector, also known as a 'genome' or 'chromosome', forms a candidate solution to the multi- dimensional optimization problem.

The initial population (at time t = 0) is chosen randomly and should be representative of as much of the search space as possible. Subsequent generations in DE can be represented by discrete time steps: t = 1, 2, ..., n etc. Since the parameter vectors are likely to be changed over different generations the following notation has been adopted here for representing the i-th vector of the population at the current generation (at time t):

$$\vec{X}_i(t) = [x_{i,1}(t), x_{i,2}(t), x_{i,3}(t).....x_{i,D}(t)]$$

For each parameter of the problem, there may be a certain range within which the value of the parameter must lie. At the beginning of a DE run, problem parameters or independent variables are initialized somewhere in their feasible numerical range. So, if the j-th parameter of the given problem has its lower and upper bound as x_i^L and x_i^U respectively, then the j-th component of the i-th population member may be initialized as

$$x_{i,j}(0) = x_j^L + rand(0,1).(x_j^U - x_j^L) \qquad (2.28)$$

where rand(0,1) is a uniformly distributed random number lying between 0 and 1.

For each individual vector $X_k(t)$ belonging to current population, DE randomly samples three other individuals $X_i(t)$, $X_j(t)$ and $X_m(t)$ from the same generation (for distinct k, i, j and m), calculates the difference of the components (chromosomes) of $X_i(t)$ and $X_j(t)$, scales it by a scalar R (ϵ [0,1]) and creates a trial offspring vector by adding the result to the chromosomes of $X_m(t)$. Thus for the n-th component of each parameter vector, we have

$$U_{k,n}(t+1) = \begin{cases} X_{m,n}(t) + R.(X_{i,n}(t) - X_{j,n}(t)) & if\ rand_n(0,1) < CR \\ X_{k,n}(t) & otherwise \end{cases} \qquad (2.29)$$

where CR ($\epsilon[0,1]$) is the crossover constant. This scheme is illustrated in Fig. 2.13.

To keep the population size constant over subsequent generations, the next step of the algorithm calls for 'selection' to determine which one between the parent and child will survive in the next generation (i.e. at time t+1). DE uses the Darwinian principle of "survival of the fittest" in its selection process which may be expressed as

$$\begin{aligned} \overrightarrow{X_i}(t+1) &= \overrightarrow{U_i}(t+1) \quad if \quad f(\overrightarrow{U_i}(t+1)) < f(\overrightarrow{X_i}(t)) \\ &= \overrightarrow{X_i}(t) \qquad if \quad f(\overrightarrow{U_i}(t+1)) > f(\overrightarrow{X_i}(t)) \end{aligned} \right\} \qquad (2.30)$$

where f(.) is the function to be minimized. If the new offspring yields a better value of the fitness function, it replaces its parent in the next generation; otherwise the parent is retained in the population. Hence the population either gets better (with respect to the fitness values) or remains the same but never deteriorates.

The DE algorithm is outlined below:

Procedure Differential-evolution
Begin
      Initialize population;
      Evaluate fitness;
      For i=0 to max-iteration do
    Begin
      Create Difference-Offspring;
      Evaluate fitness;
      If an offspring is better than its parent
      Then replace the parent by offspring in the next generation;
      End If;
    End For;
  End.

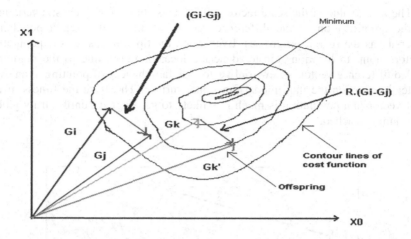

Fig. 2.13 Illustrating DE in 2-D parameter space

In the above algorithm, population is at first initialized to random values and fitness of each vector is judged according to some predefined cost function. The algorithm is then continued to generate population by invoking differential evolution and replacing parents by more fit offspring. The algorithm terminates when the fitness of the best genome is greater than a predefined value or maximum number of iterations has been attained.

2.4.2.1 Variants of Classical Differential Evolution

Generally in population-based search and optimization methods, considerably high diversity is necessary during the early part of the search to utilize the full range of the search space. On the other hand during the latter part of the search, when the algorithm is converging to the optimal solution, fine-tuning is important to locate the global optimum efficiently. Considering these issues, two new strategies [7] were proposed to improve the performance of the DE.

2.4.2.2 DERANDSF (DE with Random Scale Factor)

In the original DE [2] the difference vector $(\vec{X}_{r2}(t) - \vec{X}_{r3}(t))$ is scaled by a constant factor 'R'. The usual choice for this control parameter is a number between 0.4 and 1. We propose to vary this scale factor in a random manner in the range (0.5, 1) by using the relation

$$R = 0.5 * (1 + rand(0,1)) \tag{2.31}$$

where rand (0, 1) is a uniformly distributed random number within the range [0, 1]. We call this scheme DERANDSF (Differential Evolution with Random Scale Factor).

The mean value of the scale factor is 0.75. This allows for stochastic variations in the amplification of the difference vector and thus helps retain population diversity as the search progresses. Even when the tips of most of the population vectors point to locations clustered near a local optimum due to the randomly scaled difference vector, a new trial vector has fair chances of pointing at an even better location on the multimodal functional surface. Therefore the fitness of the best vector in a population is much less likely to get stagnant until a truly global optimum is reached.

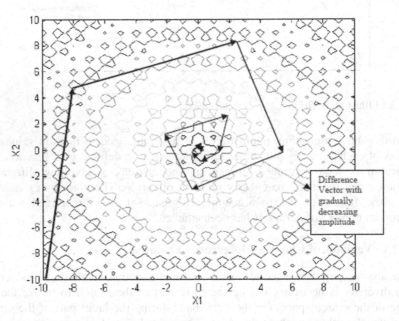

Fig. 2.14 Illustrating DETVSF scheme on two dimensional cost contours OF Ackley function

2.4.2.3 DETVSF (DE with Time Varying Scale Factor)

In most population-based optimization methods (except perhaps some hybrid global-local methods) it is generally believed to be a good idea to encourage the individuals (here, the tips of the trial vectors) to sample diverse zones of the search space during the early stages of the search. During the later stages it is important to adjust the movements of trial solutions finely so that they can explore the interior of a relatively small space in which the suspected global optimum lies. To meet this objective we reduce the value of the scale factor linearly with time from a (predetermined) maximum to a (predetermined) minimum value:

$$R = (R_{\max} - R_{\min}) * (MAXIT - iter) / MAXIT \qquad (2.32)$$

where F_{max} and F_{min} are the maximum and minimum values of scale factor F, *iter* is the current iteration number and MAXIT is the maximum number of allowable iterations. The locus of the tip of the best vector in the population under this scheme may be illustrated as in Fig. 2.14.

2.5 Biogeography-Based Optimization (BBO)

Biogeography is the study of the distribution of biodiversity over space and time. It aims to analyze where organisms live, and in what abundance. Biogeography theory grew out of the work of Alfred Wallace [8] and Charles Darwin [9]. This gives rise to an interest in the distribution of organisms. The development of biogeography allowed scientists to test theories about the origin and dispersal of populations, which spurred its application in the field of the engineering. Just as what has happened in the past few years with the areas of computer intelligence [10, 11, 12], including genetic algorithms (GAs) [13, 14, 15], ant colony optimization (ACO) [16, 17, 18, 19], particle swarm optimization (PSO) [20, 21, 22], biogeography-based optimization (BBO) as a new type of evolutionary algorithm (EA) was recently proposed. This newest EA was introduced by Simon [23] in 2008 and demonstrated good optimization performance on various benchmark functions. In the original BBO paper, it was already proven that it is competitive with other famous EAs. If its highest potential is developed and applied to more practical problems, it could become a popular EA.

When a habitat is highly populated, it has many species and thus is likely to emigrate many species to nearby habitats, while few species immigrate into it, simply by virtue of the lack of room for immigrating species. In the same way, when a habitat is sparsely populated, it has few species and thus is likely to receive many immigrants, while only a few species emigrate because of their sparse populations.

The issue of whether or not those immigrants can survive after migration is another question, but the immigration of new species can raise the biological diversity of a habitat and thereby improve the habitat's suitability for other species. At least to this point, biogeography is a positive feedback phenomenon, and we regard this phenomenon of biogeography as an optimization process. This view of the environment as an optimizing system was suggested as early as 1990s [24]. In particular, some people maintain the view that "biogeography based on optimizing environmental conditions for biotic activity seems more appropriate than a definition based on homeostasis" [25]. In fact, there are many examples of the optimality of biogeographical processes to support this view, such as the Amazon rainforest [25] and the Krakatoa island phenomenon [26].

In another view, biogeography has often been considered as a process that enforces equilibrium in habitats. Over time, the countervailing forces of immigration and emigration result in an equilibrium level of species richness in a habitat with a large number of species. Namely, equilibrium can be seen as the point where the immigration and emigration curves intersect. The equilibrium viewpoint of biogeography was first popularized in the 1960s. Since then the equilibrium perspective has been increasingly questioned by scientists.

In a word, although the natural phenomenon of biogeographical as an optimization process has been challenged, adequate literature and ideas have been put forth to explain these challenges. It must be emphasized that optimality and equilibrium are only two different perspectives on the same phenomenon in biogeography, but this debate opens up many areas of further research for engineers.

As its name implies, BBO as a novel optimization method is based on the science of biogeography. The details of the BBO approach will be presented in the next section. Just as the mathematics of biology spurred the development of other biology-based optimization methods, we can incorporate certain behaviors of biogeography into BBO to improve its optimization performance. Some of these behaviors include the effect of geographical proximity on migration rates, nonlinear migration curves to better match nature, species populations, predator/prey relationships, the effect of varying species mobility on migration rates, directional momentum during migration, the effect of habitat area and isolation on migration rates, and many others.

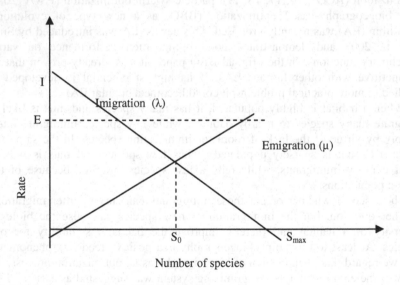

Fig. 2.15 Species model of a single habitat

The model of species abundance in a single habitat is shown in Fig. 2.15. The immigration rate λ and the emigration rate μ are functions of the number of species in the habitat. For the immigration curve, the maximum possible immigration rate to the habitat is I, which occurs when there are zero species in the habitat. As the number of species increases, the habitat becomes more crowded, fewer species are able to successfully survive immigration to the habitat, and the immigration rate decreases. The largest possible number of species that the habitat can support is S_{max}, at which point the immigration rate becomes zero.

For the emigration curve if there are no species in the habitat then the emigration rate must be zero. As the number of species increases, the habitat becomes more crowded; more species are able to leave the habitat to explore other possible residences, and the emigration rate increases. The maximum emigration rate is E, which occurs when the habitat contains the largest number of species that it can support.

The equilibrium number of species is S_0, at which point the immigration and emigration rates are equal. However, there may be occasional excursions from due to temporal effects. Positive excursions could be due to a sudden spurt of immigration (caused, perhaps, by an unusually large piece of flotsam arriving from a neighboring habitat), or a sudden burst of speciation (like a miniature Cambrian explosion). Negative excursions from could be due to disease, the introduction of an especially ravenous predator, or some other natural catastrophe. It can take a long time in nature for species counts to reach equilibrium after a major perturbation.

The immigration and emigration curves in shown in Fig. 2.16 as straight lines but, in general, they might be more complicated curves. Now, the probability Ps is the habitat contains exactly S species. Ps changes from time t to time (t + Δ t) as follows

$$P_s(t+\Delta t) = P_s(t)(1-\lambda_s\Delta t - \mu_s\Delta t) + P_{s-1}\lambda_{s-1}\Delta t + P_{s+1}\mu_{s+1}\Delta t \qquad (2.33)$$

where λ_s and μ_s are the immigration and emigration rates when there are S species in the habitat. This equation holds because in order to have S species at $(t + \Delta t)$ time , one of the following conditions must hold:

- There were S species at time t, and no immigration or emigration occurred between t and $(t + \Delta t)$;
- There were $(S - 1)$ species at time t, and one species immigrated;
- There were $(S + 1)$ species at time , and one species emigrated.

It is assumed that Δ t is small enough so that the probability of more than one immigration or emigration can be ignored.

Taking the limit of (1) as Δ t → 0 gives equation (2) shown as follows:

$$\hat{P}_s = \begin{cases} -(\lambda_s+\mu_s)P_s+\mu_{s+1}P_{s+1} & S=0 \\ -(\lambda_s+\mu_s)P_s+\lambda_{s-1}P_{s-1}+\mu_{s+1}P_{s+1} & 1\leq S\leq S_{max}-1 \\ -(\lambda_s+\mu_s)P_s+\lambda_{s-1}P_{s-1} & S=S_{max} \end{cases} \qquad (2.34)$$

Say, $n = S_{max}$ and $P = [P_0 P_1 P_2........P_n]^T$

Now, we can arrange the equations of equation (2) into the single matrix equation

$$\overset{o}{P} = AP \qquad (2.35)$$

where the matrix A is given in the following equation:

$$A = \begin{bmatrix} -(\lambda_0 + \mu_0) & \mu_1 & 0 & \cdots & & 0 \\ \lambda_0 & -(\lambda_1 + \mu_1) & \mu_2 & & \cdots & \\ \cdots & \cdots & \cdots & \cdots & \cdots & \\ \cdots & \cdots & \lambda_{n-2} & -(\lambda_{n-1} + \mu_{n-1}) & \mu_n & \\ 0 & \cdots & 0 & \lambda_{n-1} & -(\lambda_n + \mu_n) \end{bmatrix} \qquad (2.36)$$

For the straight-line curves shown in Fig. 1, we have

$$\left. \begin{array}{c} \mu_k = \dfrac{Ek}{n} \\[2mm] \lambda_k = I\left(1 - \dfrac{k}{n}\right) \end{array} \right\} \qquad (2.37)$$

Now for special case E= I , then

$$\lambda_k + \mu_k = E \qquad (2.38)$$

According to the simplified form stated in equation (6), the species model will be the following type.

2.5.1 Migration

Suppose that we have a problem and a population of candidate solutions that can be represented as vectors of integers. Each integer in the solution vector is considered to be an SIV. The assessment for the goodness of the solutions has to be done. The solutions that are good are considered to be habitats with a high Habitat Suitability Index (HIS), and those that are poor are considered to be habitats with a low HSI. HSI is analogous to "fitness" in other population-based optimization algorithms (GAs, for example).

High HSI solutions represent habitats with many species, and low HSI solutions represent habitats with few species. The identical species curve (E = I) is considered for simplicity but the S value represented by the solution depends on its HSI. S_1 in Fig. 2.16 represents a low HSI solution, while S_2 represents a high HSI solution. S_1 in Fig.2.16 represents a habitat with only a few species, while S_2 represents a habitat with many species.

The immigration rate λ_1 for S_1 will be higher than the immigration rate λ_2 for S_2. The emigration rate $\mu 1$ for S_1 will be lower than the emigration rate $\mu 2$ for S_2.

The emigration and immigration rates of each solution probabilistically share information between habitats. With probability P_{mod}, each solution is modified based on other solutions. If a given solution is selected to be modified, then the immigration rate λ to probabilistically decide whether or not to modify each suitability index variable (SIV) in that solution.

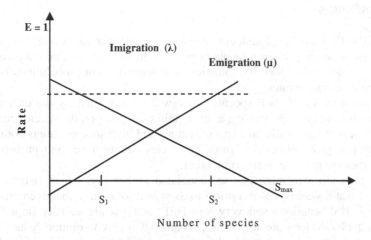

Fig. 2.16 S_1 is relatively a poor solution and S_2 relatively a good solution

If a given SIV in a given solution S_i selected to be modified, then the emigration rates μ of the other solutions to probabilistically decide which of the solutions should migrate a randomly selected SIV to solution S_i.

The BBO migration strategy is similar to the global recombination approach of the breeder GA and evolutionary strategies in which many parents can contribute to a single offspring, but it differs in at least one important aspect. In evolutionary strategies, global recombination is used to create new solutions, while BBO migration is used to change existing solutions. Global recombination in evolutionary strategy is a reproductive process, while migration in BBO is an adaptive process; it is used to modify existing islands.

To retain the best solutions in the population, some sort of elitism is incorporated. This prevents the best solutions from being corrupted by immigration.

2.5.2 Migration Algorithm

Habitat modification can loosely be described as follows:
Select H_i with probability proportional to λ_i
    If H_i is selected
        For j=1 to n
            Select H_j with probability proportional to μ_j
                If H_j is selected
                    Randomly select an SIV from H_j
                    Replace a random SIV in with
        end
      end
  end

2.5.3   Mutation

A habitat's HSI can change suddenly due to apparently random events (unusually large flotsam arriving from a neighboring habitat, disease, natural catastrophes, etc.) The model of BBO as SIV mutation, and species count probabilities is used to determine mutation rates.

The probabilities of each species count will be governed by the differential equation given in 2.39. By looking at the equilibrium point on the species curve of Fig. 2.16, it is observed that low species counts and high species counts both have relatively low probabilities and medium species counts have high probabilities because they are near the equilibrium point.

Each population member has an associated probability, which indicates the likelihood that it was expected *a priori* to exist as a solution to the given problem. Very high HSI solutions and very low HSI solutions are equally improbable. Medium HIS solutions are relatively probable. If a given solution S has a low probability P_s, then it is surprising that it exists as a solution. It is, therefore, likely to mutate to some other solution. Conversely, a solution with a high probability is less likely to mutate to a different solution. The mutation rate that is inversely proportional to the solution probability,

$$m_i = m_{max}\left(1 - \frac{P_i}{P_{max}}\right) \tag{2.39}$$

where,

$\qquad\qquad m_{max}$ is a user-defined parameter,

and $\qquad P_{max} = argmax\, P_i\,,\, i = 1,...NP$.

This mutation scheme tends to increase diversity among the population. Without this modification, the highly probable solutions will tend to be more dominant in the population. This mutation approach makes low HSI solutions likely to mutate, which gives them a chance of improving. It also makes high HSI solutions likely to mutate, which gives them a chance of improving even more than they already have. Note that we use an elitism approach to save the features of the habitat that has the best solution in the BBO process, so even if mutation ruins its HSI, we have saved it and can revert back to it if needed. So, we use mutation (a high risk process) on both poor solutions and good solutions. Those solutions that are average are hopefully improving already, and so we avoid mutating them (although there is still some mutation probability, except for the most probable solution).

> **Mutation Algorithm:** *Mutation can be described as follows:*
> *For j=1 to m*
> $\qquad$*Use λ_i and μ_i to compute the probability P_i*
> $\qquad\qquad$*Select SIV $H_i(\,j\,)$ with probability proportional to P_i*
> $\qquad\qquad\qquad$*If $H_i(\,j\,)$ is selected*

                    Replace H_i (j) with a randomly generated SIV
                    END
        END

2.6 Summary

The chapter introduced fundamental techniques of computational intelligence with special reference to fuzzy sets, neuro-computing and evolutionary algorithms. Special emphasis has been given to swarm and evolutionary algorithms, in particular Biogeography Based Optimization, Particle Swarm Optimization and Differential Evolution algorithms. A brief overview is given to neural learning, particularly supervised learning. It also includes an overview on fuzzy reasoning, starting from the first principles.

References

1. Zimmermann, H.J.: Fuzzy Set Theory and Its Applications. Kluwer Academic, Dordrecht (1991)
2. Dubois, D., Prade, H.: Fuzzy Sets and Systems: Theory and Applications. Academic Press, NY (1980)
3. Kennedy, J., Eberhart, R.: Particle swarm optimization. In: Proceedings of IEEE International Conference on Neural Networks, pp. 1942–1948 (1995)
4. Storn, R., Price, K.: Differential evolution – A Simple and Efficient Heuristic for Global continuous spaces. Journal of Global Optimization 11(4), 341–359 (1997)
5. Goldberg, D.E.: Genetic Algorithms in Search, Optimization, and Machine Learning. Addison-Wesley, Reading (1989)
6. Kennedy, J., Eberhart, R.C.: Swarm Intelligence. Academic Press (2001) ISBN 1-55860-595-9
7. Das, S., Konar, A., Chakraborty, U.K.: Two Improved Differential Evolution Schemes for Faster Global Search. In: ACM-SIGEVO Proceedings of Genetic and Evolutionary Computation Conference (GECCO 2005), Washington DC (June 2005)
8. Wallace, A.: The Geographical Distribution of Animals (Two Volumes). Adamant Media Corporation, Boston (2005)
9. Darwin, C.: The Origin of Species. Gramercy, New York (1995)
10. Hanski, I., Gilpin, M.: Metapopulation Biology. Academic, New York (1997)
11. Wesche, T., Goertler, G., Hubert, W.: Modified habitat suitabilityindex model for brown trout in southeastern Wyoming. North Amer. J. Fisheries Manage. 7, 232–237 (1987)
12. Hastings, A., Higgins, K.: Persistence of transients in spatially structured models. Science 263, 1133–1136 (1994)
13. Muhlenbein, H., Schlierkamp-Voosen, D.: Predictive models for the breeder genetic algorithm: I. Continuous parameter optimization. Evol. Comput. 1, 25–49 (1993)
14. Back, T.: Evolutionary Algorithms in Theory and Practice. Oxford Univ. Press, Oxford (1996)
15. Parker, K., Melcher, K.: The modular aero-propulsion systems simulation (MAPSS) users' guide. NASA, Tech. Memo. 2004-212968 (2004)

16. Simon, D., Simon, D.L.: Kalman filter constraint switching for turbofan engine health estimation. Eur. J. Control 12, 331–343 (2006)
17. Simon, D.: Optimal State Estimation. Wiley, New York (2006)
18. Mushini, R., Simon, D.: On optimization of sensor selection for aircraft gas turbine engines. In: Proc. Int. Conf. Syst. Eng., Las Vegas, NV, pp. 9–14 (August 2005)
19. Chuan-Chong, C., Khee-Meng, K.: Principles and Techniques in Combinatorics. World Scientific, Singapore (1992)
20. Dorigo, M., Stutzle, T.: Ant Colony Optimization. MIT Press, Cambridge (2004)
21. Dorigo, M., Gambardella, L., Middendorf, M., Stutzle, T.: Special section on 'ant colony optimization'. IEEE Trans. Evol. Comput. 6(4), 317–365 (2002)
22. Blum, C.: Ant colony optimization: Introduction and recent trends. Phys. Life Reviews 2, 353–373 (2005)
23. Onwubolu, G., Babu, B.: New Optimization Techniques in Engineering. Springer, Berlin (2004)
24. Price, K., Storn, R.: Differential evolution. Dr. Dobb's Journal 22, 18–20, 22, 24, 78 (1997)
25. Storn, R.: System design by constraint adaptation and differential evolution. IEEE Trans. Evol. Comput. 3, 22–34 (1999)
26. Michalewicz, Z.: Genetic Algorithms Data Structures _ Evolution Programs. Springer, New York (1992)
27. Rumelhart, D.E., Zipser, D.: Feature discovery by competitive learning. Cognitive Science 9, 75–112 (1985)
28. Sejnowski, T.J.: Strong covariance with nonlinearly interacting neurons. J. Math Biology 4, 303–321 (1977)
29. Takeuchi, A., Amari, S.-I.: Formation of topographic maps and columnar microstructures. Biological Cybernetics 35, 63–74 (1979)
30. Yegnanarayana, B.: Artificial Neural Networks. Prentice-Hall of India, New Delhi (1988)
31. Baird, L.C., Moore, A.W.: Gradient descent for general reinforcement learning. In: Advances in Neural Information Processing Systems, vol. 11. The MIT Press (1999)
32. Bertsekas, D.P.: Dynamic Programming And Optimal Control, vol. 1 & 2. Athena Scientific, Belmont (1995b)
33. Williams, R.J.: Simple statistical gradient-following algorithms for connectionist reinforcement learning. Machine Learning 8, 229–256 (1992)
34. Kullback, S.: Information theory and statistics. John Wiley and Sons, NY (1959)
35. Meuleau, N., Dorigo, M.: Ant colony optimization and stochastic gradient descent. Artificial Life 8(2), 103–121 (2002)

Chapter 3
Call Management in a Cellular Mobile Network Using Fuzzy Comparators

The current literature on mobile communication usually considers the channel assignment and the call admission control as two independent problems. However, in practice these two problems are not fully independent. This chapter attempts to solve the complete problem by employing a fuzzy comparator, which compares the membership of two fuzzy measurement variables to take decisions about call admission, satisfying the necessary constrains of channel assignment. Two alternative approaches to handle the problem are addressed. The first approach is concerned with the development of a fuzzy to binary mapping of the measurement variables to decision variables. The latter approach deals with fuzzy to fuzzy mapping, and then employs a fuzzy threshold to transform the fuzzy decisions into binary values for execution. A performance of both the call management techniques are studied with the standard Philadelphia benchmark and the results outperform reported results on independent call admission and channel assignment problems. The results further envisage that the latter approach is better than the former with respect to resource utilization, adaptability to the network conditions and insensitivity to load variations.

3.1 Introduction

With the rapid increase in the utilization of mobile cellular network, the call management has become an important problem. Existing techniques for call management cannot fully support the necessary user-services, such as minimization of network congestion, minimization of call drop and hard handoff. This chapter however attempts to solve the above problem with an objective to minimize hard handoff [1] and call blocking and increase the call-service. The problem addressed here is unique and new as it attempts to solve two basic problems of call management, including call admission control and channel assignment as a single (and interrelated) problem. It is apparent that the two problems coexist in mobile system, but has been solved independently to minimize the complexity of the overall problem [2-17].

Informally speaking, the call admission control (CAC) is concerned with the allocation of calls waiting for service. In a wireless network, the whole service coverage is partitioned into several continuous areas, each of which is called a

S. Ghosh and A. Konar: CAC in a Mobile Cellular Network, SCI 437, pp. 95–128.
springerlink.com   © Springer-Verlag Berlin Heidelberg 2013

cell. In each cell, a Base Station (BS) performs the central role to coordinate and relay all communications. A Mobile Station (MS) may initialize a connection request (new call) to the BS (Fig. 3.1). The CAC takes necessary decisions to accept or block the request. In case of successful initial access, the MS will roam freely with the active connection. When the MS reaches the boundary of the cell with an ongoing call, it issues the handoff request to a neighbouring cell to avoid call drop. This connection request is well known as handoff call.

The CAC module is also responsible for acceptance or rejection of the handoff call connection request. Since it is more undesirable to terminate a call in progress than to block a new call connection request, higher priority is normally given to handoff call in CAC strategy.

The common bottleneck to the call management is due to the high call congestion, which causes high call blocking and channel interference. One standard approach to solve the problem of congestion is to enhance reuse of the channels, which, however, in turn increases channel interference and decreases the quality of service (QoS). The chapter, therefore, considers channel assignment problem to resolve the call management problem.

First mobile radio system was introduced by American police department in 1921, which used the bands just above the present AM radio bands. The first commercial mobile telephone service was introduced by AT&T and South-western Bell in 1947, which had the cellular concept to increase the reuse frequency of channels though the then technology, could not cope up with the concept. In mid 1960s bell system introduced Improved Mobile Telephone Services (IMTS) which improved the frequency reuse. Their cellular network was divided into a number of circular or hexagonal cells and each had a centrally located base station and a number of radio channels.

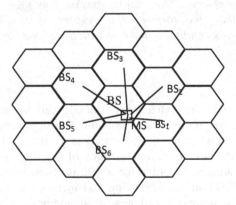

Fig. 3.1 Cell structure of wireless network

The radio channels can be a frequency, a time slot or a code sequence. The mobile devices communicates to each other with the help of a radio frequency through the base station and the base stations communicate through the Mobile Switching Centre (MSC), while MSCs communicate each through the Public Switched Telephone Networks (PSTN) shown in Fig. 3.2.

When a user starts or receives a call, he may move around in the cells of a network with the unfinished call and the network may have to handoff or transfer the call from cell to cell without the knowledge of the user. Historically, the evolution of mobile technology can be divided into 4 generations.

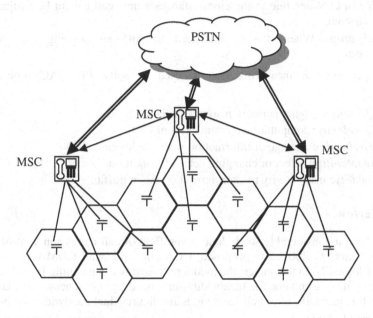

Fig. 3.2 A Typical Cellular Network

- 1G—the analog systems with major service provided was the voice transmission.
- 2G—switching to digital technology for transmission of voice and some limited data in low cost and higher capacity system.
- 3G—introduced multimedia transmission and global roaming in homogeneous network system
- 4G—extended global roaming in heterogeneous systems

Due to the increase in use and complication in technology in mobile networks, the QoS became an issue of utmost important. The Call Admission Control (CAC) mechanism is essentially needed to ensure the QoS provision by restricting the resource of the network to prevent the congestion as well as the degradation of channels currently in use. The basis concepts involved are given as follows.

1. *Channel assignment scheme:* How the calls and channels are managed by each cell using different reuse constraints [38]-[49].

2. *Handoff scheme:* How the calls are transferred from one cell to the other.

   There are two types of handoff, hard and soft [23], [27]-[37].

   a. *Hard handoff:* The old link breaks and the call drops before the new link are generated.

b.   Soft handoff: When the new channel is obtained, the call is transferred from the old channel to the new thereby causing no break in link.

3. Call block and drop: [23]-[26]

a. *Call block:* When due to the circumstances a new call cannot be assigned to a channel.

b. *Call drop:* When due to the hard handoff an ongoing call gets disconnected.

Common performance measuring criteria used to solve the CAC problem includes:

1.   *Efficiency:* higher network resource utilization
2.   *Complexity:* computational complexity of CAC
3.   *Overhead:* exchange of information with neighbour cells
4.   *Adaptability:* effect of changing network condition
5.   *Stability:* insensitivity to short term increase in traffic.

3.1.1   Review

There has been a considerable work done in the field of call admission control in recent few years. In 2005, Jue proposed a CAC in wideband CDMA network using fuzzy logic [2]. In this paper, the model proposed was a mapping from fuzzy to crisp. The fuzzy estimator for bandwidth introduced by this scheme calculates the intra-cell bandwidth, inter-cell bandwidth and the residual bandwidth for both home and neighbour cells.

First, it [6] assumes that velocity has only two values, which, however, is not realistic. Second, it considers that the acceptance or rejection of calls solely depends on the bandwidth. Third, reuse of the already used channels is not taken care of in turn compromising the efficiency. Further, computing the bandwidths of all cells over the iterations and exchanging the information among neighbouring cells is time-consuming. Finally, the effect of change in network condition is not considered and the changing traffic load is ignored.

In 2005, another Outage-based fuzzy call admission controller with multi-user detection for WCDMA systems was proposed by [7]. Here the main concern of the fuzzy model was to estimate the signal to noise ratio. The system dose not takes care of channel reuse. Each time computing the probability makes the system computationally complex. The effect of changing load is also not handled. The mobility condition of the mobile device is also not considered.

In 2006, a Neural Fuzzy Call Admission and Rate Controller for WCDMA Cellular Systems providing Multirate Services were proposed in [8]. Here, an adaptive network based fuzzy controller system is proposed using *type1* and *type2* probabilities. The system doses not consider soft handoff and dose not guarantee QoS under heavy traffic. The mobility condition of the mobile device is also not considered.

In 2006, another performance analysis of call admission control in WCDMA System with Adaptive Multi Class Traffic based on Fuzzy Logic was proposed [9]. Here the quality of service and interference is not taken care of properly.

3.1.2 Significance of the Work

The call admission control problem, addressed here, is realized by two alternative schemes shown in Fig. 3.3. Both the schemes include a fuzzy logic module for decision making about the call rejection/acceptance and also consider the channel allocation for the accepted call. The call admission decision is undertaken by considering the present dynamic scenario of the allocation matrix defined as

$$A = [f_{i,j}]$$

where, $f_{ij} = 1$, if the j-th channel in i-th cell is assigned to a call,
= 0, otherwise.

The feasibility of channels for allocation is determined by an additional channel allocation module with the consideration of compatibility matrix defined as

$$C = C_{j,k} = \begin{cases} 4, & \text{satisfying cosite constraint between j and k channel in came cell} \\ 3, & \text{satisfying adjacent channel constraint between j and k channels} \\ 2, & \text{satisfying co-channel constraint between j and k channels} \\ 0, & \text{no interferance between channel j and k} \end{cases}$$

Additional input parameters of the call admission control include hotness (degree of congestion) velocity of the mobile and its distance from the base station. It is noteworthy that in conventional call management, the feedback module through channel assignment is not considered.

The first approach undertaken to solve the problem considers a modelling from fuzzy measurements to binary decisions about call admission. The second approach is concerned with a mapping of fuzzy measurement to fuzzy control decisions [51] about call admission using Takagi-Sugano (T-S) fuzzy logic [18].

A study with the standard Philadelphia benchmark with 21 cells confirms that the proposed method outperforms most, if not all, techniques for call management and channel assignment uniquely. Summarizing, the main features of the work are listed below:

1. Minimization of call block or existing call drop, ensuring the fidelity of the network.
2. Usage of efficient channel assignment strategy for minimization of interference and ensuring quality of service.
3. Considering and solving the real world situation, where the mobile stations are moving and hence creating ambiguity in their location.
4. Reduction in the level of congestion.
5. Creating an effective MS to BS as well as global environment for better mutual understanding between the cells and enhancement of overall performance.
6. The proposed scheme utilizes both cell occupancy and mobility, and hence is expected to be more efficient.
7. The schemes are uniform cell based, and hence simple in nature.

Decision time is proactive and fast, based on a priory and feedback parameters.

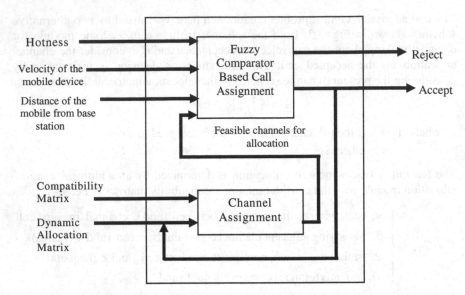

Fig. 3.3 The proposed call management

3.2   Rule Based Call Management

The chapter employs two distinct kind of fuzzy rules for decision making in automated call management in cellular network. The first set of rules maps fuzzy measurements into binary decisions about call management. The procedure applied for fuzzy decision making is different from the classical fuzzy logic. It is needless to mention here that classical Mamdani based fuzzy reasoning, maps fuzzy measurements onto fuzzy conclusions using fuzzy relations. In the present context, however, the decisions are derived based on condition checking of fuzzy linguistic variables as stated in the antecedent part of the fuzzy logic [18-20].

The second category of rules, however, is similar to classical Takagi- Sugano type, which includes an evaluation of the blocking, assignment, dropping or soft handoff membership of a call in a given cell. A fuzzy threshold is then used to transform the fuzzy decisions into binary decisions. Typically, in our experiment we presume the threshold to be 0.5; consequently when the membership obtained by firing the rules exceeds 0.5, the decision is given in favour of the fuzzy linguistic decision variable. If the consequent parts of the rule include blocking of a call, and the membership of call blocking exceeds 0.5, the call is blocked. The extension of classical fuzzy logic to derive binary decisions using thresholds is referred to as fuzzy threshold logic in this chapter.

3.2.1 Fuzzy to Binary Decision Rule Based Call Management

In this section, we briefly outline the principle of fuzzy to binary mapping using a set of rules with fuzzy propositions in the antecedent clause, and binary decision variable in the consequent part. Such policy of adoption is needed to improve system performance without increasing complexity of the reasoning algorithm.

Let $x_1, x_2, \ldots\ldots x_n$ and $y_1, y_2, \ldots\ldots y_n$ be fuzzy linguistic variable, and $\mu_{A_j}(x_j)$ and $\mu_{B_j}(y_j)$ be the membership of the variable x_i and y_i to lie in the set A_i and B_i respectively. Let d_k be a decision variable, which can hold two values: true or false. It is needless to mention that $\mu_{A_j}(x_j)$ and $\mu_{B_j}(y_j)$ lie in [0, 1]. The general form of the fuzzy rule we used are given below.

Type 1 rule:

$$IF \; \mu_{A_i}(x_i) < \mu_{B_i}(x_i) \; AND \mu_{A_k}(y_k) < \mu_{B_k}(y_k) \; AND \; ....$$

$$THEN \; d_k \; is \; true.$$

Type 2 rule:

$$IF((\mu_{A_i}(x_i) \; OR(\mu_{A_j}(x_j)) > (\mu_{A_k}(x_k)) \; AND \; (\mu_{A_j}(y_j) > \mu_{B_j}(y_j))$$

$$THEN \; d_k \; is \; false.$$

The inequality (<) in type 1 rule could be reversed (>). The OR in type 2 rule means the maximum of the membership. For example, $\mu_{A_j}(x_j) \, OR \; \mu_{B_j}(y_j)$ is same as Max[$\mu_{A_j}(x_j)$, $\mu_{B_j}(y_j)$]. The decision d_k should be true or false. The parameters defined below will be used in the rest of the chapter.

Definition 1: Hotness is defined here as the number of calls waiting to be serviced at a given time instance. If total no. of incoming calls in cell$_i$ is N$_i$, and T$_i$ is the total time of calls in cell$_i$ being serviced, then hotness of cell$_i$, denoted by hot$_i$, is measured by

$$hot_i = \frac{N_i}{T_i}.$$

Definition 2: Availability is defined here as number of free channels in a cell i. Given $f_{i,j} = 1$, if the jth channel in cell i is free. Then the availability of cell i denoted as avl_i, is measured by

$$avl_i = \sum_j f_{i,j}.$$

Definition 3: Feasibility is a parameter used to check the viability of channel assignment to a incoming call in a given cell. To test the feasibility of assignment of channel j to an incoming call in cell i, when the channel k of cell i is already in service, $\forall$ k, we define a measure of feasibility by $feas^j$, using the soft constraints for channel assignment indicated in [42], where

$$feas^j = \sum_k (C_{j,k} - |f_{i,j} - f_{i,k}|).$$

If $feas^j > 0$ then the j-th channel is assigned to an incoming call in cell i. Otherwise, the assignment is abandoned.

Definition 4: Velocity is defined as the velocity with which a MS is moving while in service and is denoted as *vel*.

Definition 5: Distance is defined as the distance of the moving MS from the BS and is hereafter referred to as *dist*.

3.2.2   Fuzzy Membership Evaluation

In call management, the parameters like availability, hotness, velocity of the mobile device and distance from the base station are always dynamically fluctuating with time. Construction of rules for all possible values of the parameters becomes tedious, and also not suggestive as matching of the if-part of the rules with measured value of parameters for a large no of rules takes excessive time, and thus not amenable for execution in real time.   One approach to reduce the matching time is to partition the parametric space into intervals, and check the existence of a measured parameter in intervals. Such crisp partitioning is permissible as long as a particular value of a parameter in the if-part of a rule is independent (with respect to decision) over the other parameters in the if-part of the same rule. Fortunately, this is taken care of in fuzzy sets, as it allows overlapping among the partitions in individual parametric space. The parameters: hotness, availability, velocity, and distance here are encoded (fuzzified) in three scales HIGH (hereafter HI), LOW (hereafter LO) and MEDIUM (hereafter MED), while feasibility is encoded into two fuzzy scales: HIGH and LOW. Fig. 3.4 shows the fuzzy membership curve of all the parameters stated above.

3.2.2.1   Fuzzy Rules for Call Admission

The rules used in the present context are given here with justifications below.

1. IF $(\mu_{HI}(avl_i) \ OR \ \mu_{MED}(avl_i)) > \mu_{LO}(avl_i) \ AND \ (\mu_{HI}(feas^j) > \mu_{LO}(feas^j))$

    THEN assign new call to jth channel.

2. IF $(\mu_{LO}(avl_i) \ OR \ \mu_{MID}(avl_i) > \mu_{HI}(avl_i)) \ AND$
    $(\mu_{HI}(feas^j) > \mu_{LO}(feas^j)) AND (\mu_{HI}(hot_N) > \mu_{LO}(hot_N)$
    $OR \ \mu_{MED}(hot_N)) \ AND \ (\mu_{HI}(avl_N) > \mu_{LO}(avl_N) \ OR \ \mu_{MID}(avl_N))$

THEN block the cell.

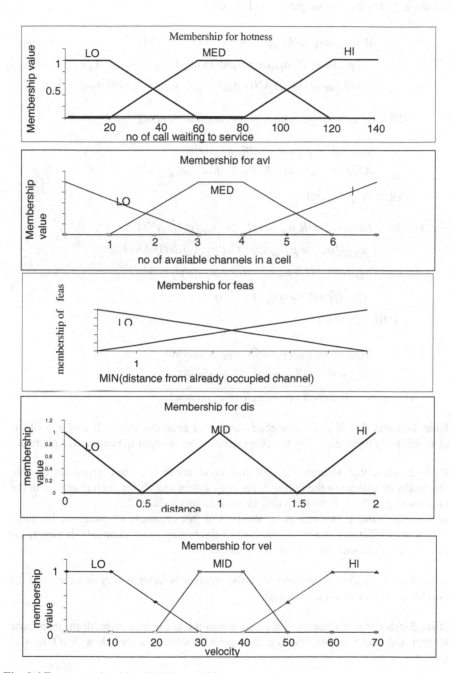

Fig. 3.4 Fuzzy membership of all the variables

In rule 2, N denotes the neighborhood of cell i

1.   IF$(\mu_{LO}(avl_i)$ OR $\mu_{MID}(avl_i) > \mu_{HI}(avl_i))$ AND
       $(\mu_{HI}(feas^j) > \mu_{LO}(feas^j))AND(\mu_{HI}(hot_N) > \mu_{LO}(hot_N)$
       OR $\mu_{MED}(hot_N))$ AND$(\mu_{HI}(avl_N) > \mu_{LO}(avl_N)$ OR $\mu_{MID}(avl_N))$

   THEN search new neighbor N and do Soft Handoff

2.   IF $(\mu_{HI}(vel) > \mu_{MID}(vel)$ OR $\mu_{LO}(vel))$
            AND $(\mu_{FAR}(dis)$ OR $\mu_{MED}(dis) > \mu_{NEAR}(dis))$

   THEN drop the call

3.   IF $(\mu_{HI}(vel)$ OR $\mu_{MID}(vel) > \mu_{LO}(vel))$ AND
            $(\mu_{MED}(dis) > \mu_{FAR}(dis)$ OR $\mu_{NEAR}(dis))$ AND
            $(\mu_{LO}(avl_i)$ OR $\mu_{MID}(avl_i) > \mu_{HI}(avl_i))$ AND
            $((\mu_{HI}(feas^j) > (\mu_{LO}(feas^j))$

   THEN do SHO

4.   IF $(\mu_{LO}(vel) > \mu_{MID}(vel)$ OR $\mu_{HI}(vel))$ AND
            $(\mu_{FAR}(dis)$ OR $\mu_{MED}(dis) < \mu_{NEAR}(dis))$

   THEN the call carries on with existing channel.

Rule 1 asserts that if a cell has good number of available channels and any of the channels has a high feasibility to accept the call then assigns the call to that channel.

Rule 2 states that though the cell has good number of free channels but the feasibility of assigning the call to them is low then search for such channels in the neighboring cells N. If no such cell exists then block the call.
Rule 3 states that if in a cell good numbers of free channels are available but none fits the feasibility criteria then search the neighboring cells' channels. If any of the cell has such channel do SHO to that channel.

Rule 4 states that in a situation where the receiver is moving very fast and is at the boundary of the cell drop such call.

Rule 5 states that if velocity is high to medium and distance is medium from base station and there is neighboring cell with feasible channel then SHO to that channel

Rule 6 states that if the speed is very slow and is very near to the base station, do not change the channel.

3.2.3   Fuzzy to Fuzzy Mapping for CAC

At the time of assigning a channel to a call, the most important factor to be taken care of is to ensure the QoS, satisfying the feasibility criteria. In case the channels satisfy the feasibility criteria, they are fit for new assignment. In fuzzy to binary mapping, only the fittest channel in a cell is used for new assignment. Determination of the fittest channel there has been performed by selecting the channel with the highest grade of fuzzy membership of High-feasibility set. But in case of fuzzy to fuzzy mapping, the decision variable is fuzzy and is connected with the measurement variables: feas, avl, hot, vel and dis through fuzzy relations.

In order to arrive at a binary decision from the estimated fuzzy decision variable, a threshold is considered, exceeding which the decision is considered true, and else it is regarded false. So a channel with lesser feasibility than the highest is also favoured, if the membership of the decision variable exceeds the threshold. As a result, more channels are in use at a time than in the previous mapping scheme. But choice of suitable threshold ensures that the soft constraints are not violated ensuring QoS.

The fuzzy rules in the present context are Tkagi-Sugeno type, but the reasoning mechanism we introduced here is considerably different from classical Takagi-Sugeno model [3], [4].

The general form of rules used includes

Type 1:

$$if\,(\mu_{A_i}(x_i)\;OR\;\mu_{A_j}(x_j) > \mu_{A_k}(x_k)$$

$$AND\;(\mu_{B_j}(y_j) > \mu_{B_k}(y_k)),\;\forall j$$

$$AND \quad ..\,...........\,..$$

Then d_l is with t ruth value

$$\mu_c(d_l) = max\{(\mu_{A_i}(x_i),(\mu_{A_j}(x_j))\}^* \max_j\{\mu_{B_j}(y_j)\}$$

where A_i, A_j, A_k, B_j, B_k and C are fuzzy sets in respective universe discourses. The "*" denotes product of the arguments on both side of it.

Type 2:

$$if\,(\mu_{A_i}(x_i)\;OR\;\mu_{A_j}(x_j) > \mu_{A_k}(x_k)\;AND$$

$$(\mu_{B_i}(y_i) > ((\mu_{B_j}(y_j)\;OR\;(\mu_{B_k}(y_k)),\;\forall i$$

$$OR\,((\mu_{C_j}(z_j) > \mu_{C_k}(z_k)),\;\forall j\;AND$$

$$(\mu_{A_i}^N(x_i) > ((\mu_{A_j}^N(x_j)\;OR(\mu_{A_k}^N(x_k))),\;\forall i)\;AND \qquad ....$$

Then d_l is with t ruth value

$$\mu_D(d_l) = min[1,\,max\,\{(\mu_{A_i}(x_i),(\mu_{A_j}(x_j))\}^* \max_j\{(\mu_{B_j}(y_j)\} +$$

$$+ \max_j\,\{(\mu_{C_j}(z_j)\}^* \max_i\,\{(\mu_{A_i}^N(x_i)\}]$$

where A_i, A_j, A_k, B_i, B_j, B_k, C_j, C_k, and D are fuzzy sets in respective universes. The conjunctive antecedent clauses' contribution to decision variable is realized using product (*), while the contribution of to disjunctive clauses is realized in decision variable by summation (+) operation.

The fuzzy implication rules given below use a decision variable for un-serviced call (hereafter abbreviated as uc) in five fuzzy sets ASSIGNMENT (hereafter ASSIGN), BLOCK, SOFT-HANDOFF (abbreviated as SHO), DROP, and CONTINUE (abbreviated as CONT).

3.2.4   Fuzzy Rules Used and Justification

1.   IF $(\mu_{HI}(avl_i) \text{ OR } \mu_{MID}(avl_i)) > \mu_{LO}(avl_i)$

       AND $(\mu_{HI}(feas^j) > \mu_{LO}(feas^j))$

Then assign new call to j^{th} channel with membership

$$\mu_{ASIGN}(uc) = max\{\mu_{HI}(avl_i), \mu_{MID}(avl_i)\} * \underset{j}{max}\{\mu_{HI}(feas^j)\}.$$

Membership of assignment is expressed in terms of availability and feasibility, and if the membership value exceeds a threshold of 0.5 then the channel j of cell i is assigned to the call, and rejected otherwise.

2.   IF $(\mu_{HI}(avl_i) \text{ OR } \mu_{MID}(avl_i) > \mu_{LO}(avl_i))$ AND

       $((\mu_{LO}(feas^j) > \mu_{HI}(feas^j))$ AND

       $(\mu_{LO}(hot_N) > \mu_{HI}(hot_N) \text{ OR } \mu_{MED}(hot_N))$

       THEN block the cell with membership of block as

$$\mu_{BLOCK}(uc) = min\,[1, \mu_{LO}(avl_i) * \underset{j}{max}\,\{\mu_{LO}(feas^j)\} +$$

$$\underset{N}{max}\,\{\mu_{LO}(hot_N)\}]$$

It is clear from the above expression that membership of call blocking is a function of memberships of low availability and low feasibility or low hotness, and the call is blocked if the membership of call blocking exceeds a threshold = 0.5.

3.   IF $(\mu_{LO}(avl_i) \text{ OR } \mu_{MID}(avl_i) > \mu_{HI}(avl_i))$

           AND $(\mu_{HI}(feas^j) > \mu_{LO}(feas^j))$

           OR $((\mu_{HI}(hot_N) > \mu_{LO}(hot_N) \text{ OR } \mu_{MED}(hot_N))$

           AND $(\mu_{HI}(avl_N) > \mu_{LO}(avl_N) \text{ OR } \mu_{MID}(avl_N)))$

THEN search new neighbor N and do Soft Handoff with membership given by

$$\mu_{SHO}(uc) = min[1, max\{\mu_{MID}(avl_i),\mu_{LO}(avl_i)\}* \max_{j}\{\mu_{LO}(feas^{j})\} +$$

$$\max_{N}\{\mu_{LO}(hot_N)\}* \max_{N}\{\mu_{HI}(avl_N)\}].$$

Membership of SHO is expressed in terms of inadequate availability of feasible channels in the cell and the hotness of the neighbors and availability of feasible channels in such neighbors and is accomplished if the membership value exceeds threshold=0.5

4. IF $(\mu_{HI}(vel) > \mu_{MID}(vel)$ OR $\mu_{LO}(vel))$ AND

$(\mu_{FAR}(dis)$ OR $\mu_{MED}(dis) > \mu_{NEAR}(dis))$

THEN drop the call with membership as
$$\mu_{DROP}(uc) = \mu_{HI}(vel)*max\{\mu_{FAR}(dis),\mu_{MED}(dis)\}.$$

Membership of drop is expressed in terms of high velocity and the distance away from the BS and the call is dropped when membership value exceeds a threshold=0.5.

5. IF $(\mu_{HI}(vel)$ OR $\mu_{MID}(vel) > \mu_{LO}(vel))$ AND

$(\mu_{MED}(dis) > \mu_{FAR}(dis)$ OR $\mu_{NEAR}(dis))$

OR $((\mu_{HI}(avl_N) > (\mu_{LO}(avl_N)$ OR $\mu_{MID}(avl_N)))$ AND

$((\mu_{HI}(feas^{N_j}) > (\mu_{LO}(feas^{N_j})))$

THEN do SHO with membership
$$\mu_{SHO}(uc) = min[1,max\{\mu_{HI}(vel),\mu_{MID}(vel)\}*max\{\mu_{MED}(dis)\} +$$

$$\max_{N}\{\mu_{HI}(avl_N)\}* \max_{j}\{\mu_{HI}(feas^{N_j})\}]$$

The membership of SHO is expressed in terms of higher range of velocity with medium distance of MS from BS and availability of feasible channels in near neighbor and is done if membership value exceeds the threshold =0.5.

6. IF $(\mu_{LO}(vel) > \mu_{MID}(vel)$ OR $\mu_{HI}(vel))$ AND

$(\mu_{FAR}(dis)$ OR $\mu_{MED}(dis) < \mu_{NEAR}(dis))$

THEN carry on with existing channel with membership
$$\mu_{CONT}(uc) = \mu_{LO}(vel)* \mu_{NEAR}(dis).$$

The membership CONT is expressed in terms of low velocity and nearing distance of MS from BS and is accomplished if membership value exceeds threshold=0.5.

3.3   Computer Simulation

Here we have considered two different decision schemes for assignment or rejection of a call in a channel. Though the decisions are estimated differently, the antecedents are same in both the cases. Hence the same call management strategy is used in both the case which is described below.

3.3.1   Principles of Call Management Strategy

In a network every cell contains channels, some of which are busy in servicing calls, while the rest are free. When a new call arrives in a cell, it is to be assigned to the free channels of the cell, satisfying the soft constraints of channel assignment. If the cell does not have such free channels, the neighboring cells are searched for free channels, and once found they are assigned to the calls. If no such channels are available in near neighbors (adjacent six cells) then the call is rejected.

This, however, is valid as long as the MS are static. But in case the MS starts moving during a call service, it may move out of the cell. To avoid call drop, a new channel from the current cell needs to be allocated when the MS moves out of the cell. If no such channel is found, the call is dropped. The time of search is considered and if the MS is moving out of the cell very fast the call is dropped. If it is moving slowly and is well within the range of the BS then the channel in use continues the service.

In the time of call assignment the calls generated due to SHO are given higher priority over new ones, to maintain the QoS of on-going calls. The calls which are occupying the channels for long time are forcefully dropped to free the channels for reuse. The call management strategy is proposed based on the above considerations.

3.3.2   Call Management Strategy

1. A new call arrives in the cell.
2. If it is a new call, set SHO flag to 0 else (if it is due to handoff) set to 1.
3. Check if the call is moving or static; if moving repeat from step 10.
4. Drop all the calls with call time more than T, and set all the free channels due to call drop and hang-ups available; and refresh the dynamic allocation table and hotness table.
5. Check the availability of the channels in the cell and list all the available channels with the membership of avl.
6. Check the feasibility of the available channels and find the membership of feas.
7. If membership of avl and feas both are high, then assign the call to that channel of the cell, and start counting call time (so as to measure call duration to drop long calls by strategy 4) and go to step 1.
8. If feas is not high, search the highest feasible channel in the near neighbour with highest avl, and if found do SHO; start counting call time and go to step 1.

9. If no such cell found satisfying step 6, block the call and go to step 1.
10. Find the membership of vel and dis.
11. If vel and dis both are low, the channel used for call service is retained in the cell, and go to step 1.
12. if vel is med or high and dist is med, then find the membership of avl for the neighbours, and go to step 7.
13. If both vel and dist are respectively high and far, then drop the call and go to step 1.

The membership curves of the variables are given in Fig 2.4.

3.3.3 Sample Runs for Fuzzy to Binary

The strategy of call admission has two modules (Fig. 3.5). One describes the call admission when the MS are static the other describes the same when the MS are moving. Three specific cells are taken to demonstrate the rules are described in Table. 3.1.

Table 3.1 Measurement of System Features Following the Above Definitions

No of cell	vel	dist	avl	hot	feas
7	5	1	5	55	1,2,4,5
21	55	.4	6	60	1,2,4,5,6,7
9	55	1.6	5	103	1,2,4,5,7

The first module is described in Fig. 3.6a. Here at a particular instance of time when a call arrives in the cell the membership of hotness is calculated using hotness curves. Then the membership or availability of channels in the cell as well as the membership of feasibility of the channels are calculated. Then rules are fired on the basis of membership values, and a binary decision is obtained.

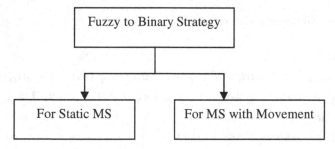

Fig.3.5 Block diagram for fuzzy to binary strategy

RULE1:

$IF(\mu_{HI}(avl_i) \; OR \; \mu_{MID}(avl_i) > \mu_{LO}(avl_i) \; AND \mu_{HI}(feas^j) > \mu_{LO}(feas^j)$
Assign new call to chanel j in cell i

RULE2:

IF $(\mu_{HI}(avl_i)$ OR $\mu_{MID}(avl_i) < \mu_{LO}(avl_i))$ AND

$(\mu_{HI}$ $(feas^{\,j}) < \mu_{LO}$ $(feas^{\,j}))$ AND$(\mu_{HI}(hot_N)$ OR $\mu_{MID}(hot_N) < \mu_{LO}(hot_N))$

Block the call

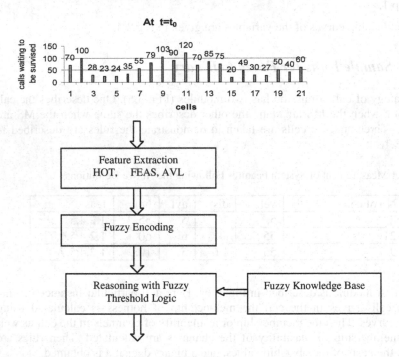

Fig. 3.6a Module describing static call service

RULE3:

IF $(\mu_{HI}(avl_i) < \mu_{MID}(avl_i)$ OR $\mu_{LO}(avl_i))$AND $(\mu_{HI}$ $(feas^{\,j}) < \mu_{LO}$ $(feas^{\,j}))$

AND $(\mu_{HI}(hot_N) > \mu_{MID}(hot_N)$OR $\mu_{LO}(hot_N))$AND $\mu_{HI}(avl_N) > \mu_{lo}(avl_N)$

OR $\mu_{MED}(avl_N)$

Search new neighbour N and do SHO

RULE 4:

IF $(\mu_{HI}(vel) > \mu_{MID}(vel)$ OR $\mu_{LO}(vel))$AND $(\mu_{FAR}(dis)$ OR $\mu_{MED}(dis) >)$

$\mu_{NEAR}(dis)$

DROP THE CALL

RULE5:

IF($\mu_{HI}(vel)$ OR $O\mu_{MID}(vel) > \mu_{LO}(vel))$ AND ($\mu_{FAR}(dis)$ OR

$\mu_{NEAR}(dis)$) $< \mu_{MED}(dis)$AND ($\mu_{LO}(avl_N)$ OR

$\mu_{MID}(avl_N)$ $\langle \mu_{HI}(avl_N)$of nbd (N)AND($\mu_{HI}(feas^j)> \mu_{LO}(feas^j)$))

DO SHO

The module in Fig. 3.6b is the second part of the scheme proposed by us. Here the situation is considered when the MS starts moving inside the cell or may move out of the cell as well. In such case the membership value of velocity of MS as well as the membership of distance from base station is also important to arrive at a decision, and hence they are also considered together.

In addition, the availability of channels in the neighbouring cell as well as feasibility of using such channels are equally important in decision making. Depending on all such membership values the rules are fired to get a binary decision for the course of action to be followed.

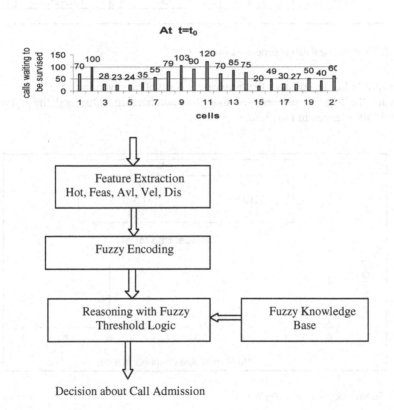

Fig. **3.6b** Module describing call service when MS in in motion

3.3.4 Numerical Examples

Let us find out how the system works in a situation where three cells say, the 9th, 11th and 5th are having static calls at an instant $t = t_0$. By calculating the membership value, we find the following facts which are shown in Fig. 3.7a.

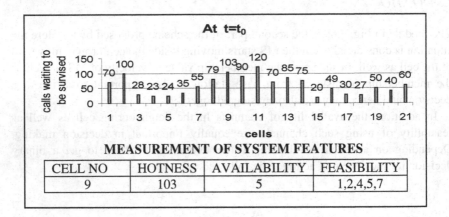

CELL NO	HOTNESS	AVAILABILITY	FEASIBILITY
9	103	5	1,2,4,5,7

Fig. 3.7 When a new call is generated in cell 9

Example 1: Let us consider the network in a situation given in Fig. 3.7

From the Fig 3.7 we can see that the membership value and the curves of availability is given in Fig. 3.7a.

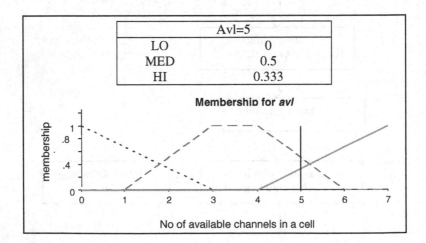

Fig. 3.7a Membership curve for avl

From Fig. 3.7a, it is evident that $\mu_{HI}(avl) > (\mu_{MED}(avl) OR \mu_{LO}(avl))$

The fuzzy membership calculation of hotness is given in Fig. 3.7b.

From Fig.3.7b it follows that $\mu_{HI}(hot) > (\mu_{MED}(hot) OR \mu_{LO}(hot))$

Then the feasibility of the free channels are considered. Fig. 3.7c below demonstrates the feasibility of all the free channels mentioned in Fig.3.2a

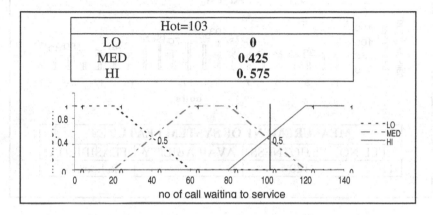

Fig. 3.7b Membership curve for hot in cell 9

From Fig. 3.7c we can conclude $\mu_{LO}(feas^{j}) > \mu_{HI}(feas^{j}) \forall j \in \{1,2,4,5,7\}$

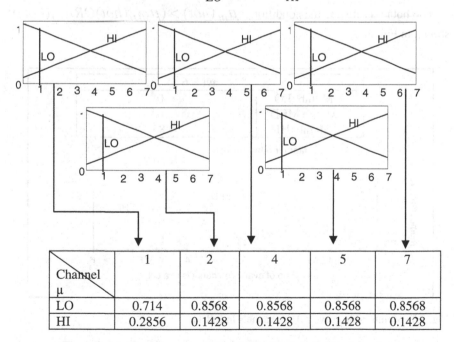

Channel μ	1	2	4	5	7
LO	0.714	0.8568	0.8568	0.8568	0.8568
HI	0.2856	0.1428	0.1428	0.1428	0.1428

Fig. 3.7c Membership curve for feasibility of all available channels

So in the given condition stated above when a new call is generated in cell 9 then all the conditions trigger Rule 2 and the call is blocked. All the tables hereafter are refreshed.

Now consider cell 11 given in Fig. 3.7d

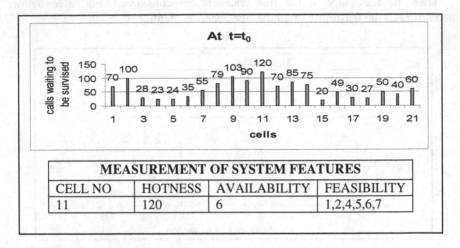

MEASUREMENT OF SYSTEM FEATURES			
CELL NO	HOTNESS	AVAILABILITY	FEASIBILITY
11	120	6	1,2,4,5,6,7

Fig. 3.7d When a new call is generated in cell 11

It is evident from Fig. 3.7e   that $\mu_{MED}(avl) > (\mu_{HI}(avl)OR\mu_{LO}(avl))$

The hotness satisfies the condition $\mu_{HI}(hot) > (\mu_{MED}(hot)OR\mu_{LO}(hot))$ shown in Fig. 3.7f.

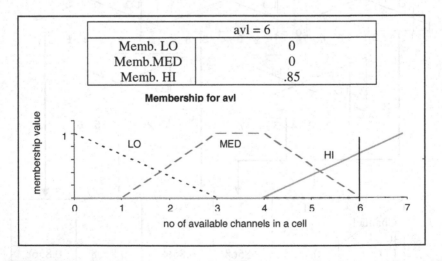

avl = 6	
Memb. LO	0
Memb.MED	0
Memb. HI	.85

Fig. 3.7e Membership curve avl of channels in cell 11

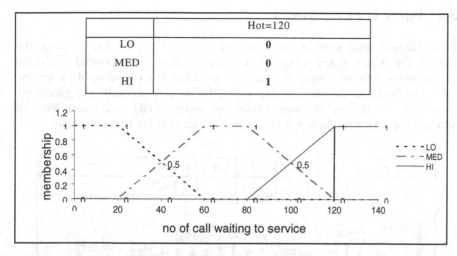

Fig. 3.7f Membership for hotness =120

The   feasibility   as   shown   in   the   Fig.   3.7g   shows   that
$\mu_{HI}(feas^6) > \mu_{LO}(feas^6)$ and for other j $\mu_{LO}(feas^j) > \mu_{HI}(feas^j)$.

So when a new call arrives in cell 11 the 6th channel satisfies all the conditions
of rule. Hence the new call is assigned to the 6th channel of cell 11.

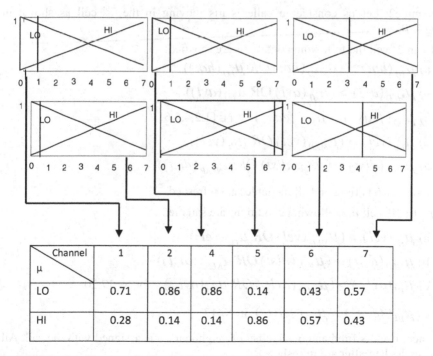

Channel μ	1	2	4	5	6	7
LO	0.71	0.86	0.86	0.14	0.43	0.57
HI	0.28	0.14	0.14	0.86	0.57	0.43

Fig. 3.7g Membership curve for feasibility of all available channels in cell 11

3.4   Fuzzy to Fuzzy Model

In the fuzzy to fuzzy scheme, the 1st module given in Fig. 3.6a, 2.6b is almost the same as the module given in Fig. 3.8. The notable change incorporated here is the output which is fuzzy instead of binary. Hence depending on the input values, the rules are fired and the membership values for un-serviced calls are generated. Depending on the membership values for assign, SHO and reject the call admission decision is taken. A thresholding is done to arrive to the decision.

No of cell	vel	dist	avl	hot	feas
7	5	1	5	55	1,2,4,5

At $t=t_0$

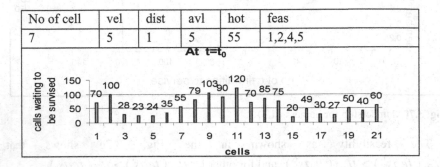

Fig. 3.7h Membership for cell 7

Example2: Let us consider a caller starts moving in the 7th cell as shown in Fig.3.7h.

1) In 7th cell the following conditions are satisfied

 a) $\mu_{HI}(hot) > (\mu_{MED}(hot) \, OR \, \mu_{LO}(hot))$

 b) $\mu_{MED}(avl) > (\mu_{HI}(avl) \, OR \, \mu_{LO}(avl))$

 c) $\mu_{LO}(feas^j) > \mu_{HI}(feas^j) \quad \forall \, j \in \{1,2,4,5,6\}$

 d) $\mu_{LO}(vel) > (\mu_{MED}(vel) \, OR \, \mu_{HI}(vel))$

 e) $\mu_{MED}(dist) > (\mu_{FAR}(dist) \, OR \, \mu_{NEAR}(dist))$

Hence rule 6 is fired and all the tables are refreshed.

2) In 21st cell the following conditions are satisfied

 a) $\mu_{HI}(vel) > (\mu_{MED}(vel) \, OR \, \mu_{LO}(vel))$

 b) $\mu_{MED}(dist) > (\mu_{FAR}(dist) \, OR \, \mu_{NEAR}(dist))$

 c) $\mu_{HI}(avl_N) > (\mu_{MED}(avl_N) \, OR \, \mu_{LO}(avl_N))$ for N=17,20

 d) $\mu_{HI}(feas^6) > \mu_{LO}(feas^6)$ in N=17

Hence rule 5 is fired and 6th channel of neighbour 17 is assigned with the call. All the tables hereafter are refreshed.

3) In 9th cell the following conditions are satisfied

a) $\mu_{HI}(vel) > (\mu_{MED}(vel)\ OR\ \mu_{LO}(vel))$

b) $\mu_{FAR}(dist) > (\mu_{MED}(dist)\ OR\ \mu_{NEAR}(dist))$

Hence rule 4 is fired and call is dropped.

The Table 3.2 Shows all the membership values of all variable in the three given cells.

Table 3.2 Membership Value for all the variables when MS is moving

Cell No	7	21	9
Hot	$\mu_{LO}=0$ $\mu_{MED}=.425$ $\mu_{HI}=.575$	$\mu_{LO}=0$ $\mu_{MED}=.425$ $\mu_{HI}=.575$	$\mu_{LO}=0$ $\mu_{MED}=.425$ $\mu_{HI}=.575$
Avl	$\mu_{LO}=0$ $\mu_{MED}=.5$ $\mu_{HI}=.333$	$\mu_{LO}=0$ $\mu_{MED}=0$ $\mu_{HI}=1$	$\mu_{LO}=0$ $\mu_{MED}=.5$ $\mu_{HI}=.333$
Vel	$\mu_{LO}=1$ $\mu_{MED}=0$ $\mu_{HI}=0$	$\mu_{LO}=0$ $\mu_{MED}=.25$ $\mu_{HI}=0.75$	$\mu_{LO}=0$ $\mu_{MED}=.25$ $\mu_{HI}=.7$
Dist	$\mu_{LO}=0$ $\mu_{MID}=1$ $\mu_{FAR}=0$	$\mu_{LO}=.2$ $\mu_{MID}=0$ $\mu_{FAR}=0$	$\mu_{LO}=0$ $\mu_{MID}=0$ $\mu_{FAR}=.2$
feas	$\mu^1_{LO}=.714$ $\mu^1_{HI}=.2856$ $\mu^2_{LO}=.8568$ $\mu^2_{HI}=.1428$ $\mu^4_{LO}=.8568$ $\mu^4_{HI}=.1428$ $\mu^5_{LO}=.714$ $\mu^5_{HI}=.2856$	$\mu^1_{LO}=.714$ $\mu^1_{HI}=.2856$ $\mu^2_{LO}=.8568$ $\mu^2_{HI}=.1428$ $\mu^4_{LO}=.8568$ $\mu^4_{HI}=.1428$ $\mu^5_{LO}=.714$ $\mu^5_{HI}=.2856$ $\mu^6_{LO}=.4284$ $\mu^6_{HI}=.5712$ $\mu^7_{LO}=.5712$ $\mu^7_{HI}=.4284$	$\mu^1_{LO}=.714$ $\mu^1_{HI}=.2856$ $\mu^2_{LO}=.8568$ $\mu^2_{HI}=.1428$ $\mu^4_{LO}=.8568$ $\mu^4_{HI}=.1428$ $\mu^5_{LO}=.8568$ $\mu^5_{HI}=.1428$ $\mu^7_{LO}=.8568$ $\mu^7_{HI}=.1428$ $\mu^1_{LO}=.714$ $\mu^1_{HI}=.2856$

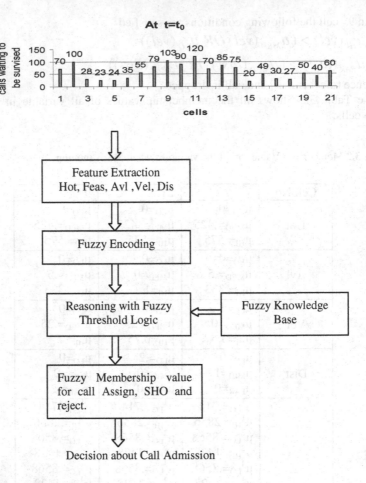

Fig. 3.8 Module describing call service for fuzzy to fuzzy model

3.5 Experiments and Simulation

For simulation we have considered a situation where

1. No. of cells = 21
2. no of channels in each cell= 7
3. Minimum reuse distance in a cell(cosite) $C_{i,i}$=4
4. Co channel constraint $C_{i,m}$=2
5. Adjacent Channel constraint $C_{i,m}$=3
6. minimum call arrival in a system= 21
7. maximum call arrival in a system = 150
8. base station separation= 2 km
9. minimum call hang-up in a cell =1
10. maximum time before call drop in a hot cell= 1 hr
11. maximum velocity= 70 km/hr(city traffic)

At $t = t_0$, 27 calls were serviced and 4 calls were rejected. But the change of hotness may be different due to the hang up of calls by the users. Fig. 3.9b. shows a sample run and how the hotness changes and calls getting serviced in the processes. It is seen that the number of calls serviced increases steadily over the time.

3.6 Results and Interpretation

We define call assignment and call rejection probability to measure the performance of our proposed system of call management.
 Let

 N_a be the number of calls assigned in a cell,
 N_s be the number of incoming calls waiting to be serviced and
 N_r be the number of calls rejected.

Then we define call assignment probability and call rejection probability by the following expression

$$P_A = P \text{ (call assign)} = \frac{N_a}{N_s} \quad , \quad P_R = P \text{ (call rejection)} = \frac{N_r}{N_s}$$

We first study the performance of the fuzzy to binary and fuzzy to fuzzy algorithm separately.

3.6.1 Performance of Fuzzy to Binary Strategy

In Fig. 3.9 we present a graphical representation of the no. of serviced calls and rejected calls against the total incoming calls. It is apparent from Fig. 3.9 that the system starts with a very few incoming calls, i.e. N_s is low. Naturally, plenty of channels remain free for further call assignment in the cells. Consequently, the call assignment continues to the free channels, and resulting in a significant rise in the number of call assigned (serviced) N_a rises with N_s, when N_r remains constant.

 But as Ns increases and the channels get allocated to the calls, free and assignable channels gradually decrease. This happens when N_s exceeds 50(approx) (Fig. 3.9). Then the system experiences congestion and N_r gradually starts increasing whereas N_a gets flatten. Under such situation, the call management scheme starts executing a call drop strategy, where the channels in service for more than a given amount of time are set free. The channels for the normal hang-up (terminated) calls are also set free for subsequent reuse. And the call allocation matrix is refreshed. This dropping and reuse job is repeated in regular intervals. And here after N_a goes on increasing with N_s and N_r flattens.

 Table 3.3 provides a situation of network after running Fuzzy to Fuzzy CAC Algorithm

Table 3.3 Call Management in a Network

CELL NO.	Call Blocked And Dropped	Soft Handoff	Call Serviced
1		1	
2		1	
3	1		
4		1	
5	1		
6	1		
7		1	
8		1	
9		1	
10			1
11			1
12			1
13		1	
14		1	
15		1	
16			1
17		1	
18	1		
19			1
20			1
21			1
Total	4	10	7

Fig. 3.10 provides a graphical representation of the probability of call assignment P_A and probability of call rejection P_R with increasing number of calls waiting for being serviced N_s. Initially the P_A grows and P_R falls sharply since there are plenty of free channels in the system. Then as N_s increases, the congestion grows and calls start dropping. So P_A starts dropping, and P_R starts growing up. At that point the call management strategy starts dropping long calls repetitively after certain interval of time and reuses the free channels.

Hence, P_A moves up and stays up since the dropping and reuse of channels are repeated in regular intervals. But the slope of P_A flattens up when Ns goes beyond a certain stage (approx 110). This signifies that with limited no. of channels, the configuration indicated in Fig. 3.1, has no more free channels, and thus cannot serve the excessive demand. One way to meet this demand, is to have adequate channels for a given system configuration, so that channels are free even when the call demand is height.

Table 3.4 shows the run of Fuzzy to Fuzzy system in a regular interval.

Table 3.4 Performance of the scheme after certain interval of time

HOTNESS	CALL SERVICED	CALL REJECTED
Hotness at t=t0	27	4
Hotness at t=t4	29	6
Hotness at t=t9	34	6
Hotness at t=t14	38	6

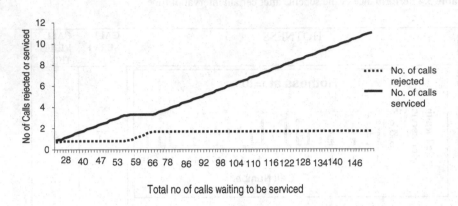

Fig. 3.9 Total number of call dropped and serviced over the time

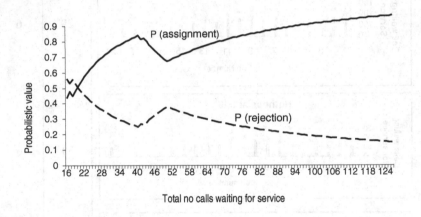

Fig. 3.10 The call- drop probability and call assign probability

3.6.2   *Performance of Fuzzy to Fuzzy Strategy*

To study the performance of fuzzy to fuzzy decision making in call management strategy, we again have plotted N_a and N_r against N_s in Fig. 3.11. We also plot the P_A and P_R against N_s in Fig. 3.12.

In Fig. 3.11 we see that the congestion is reached much before (around 45) that it was in fuzzy to binary scheme. This is because the fuzzy to fuzzy scheme takes more channels as available for allocating calls.

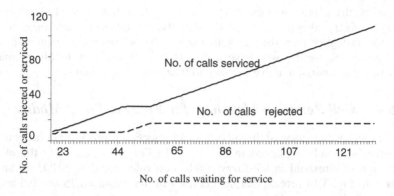

Fig. 3.11 Total number of call dropped and serviced over the time

Fig. 3.12 shows that P_A always stays higher than P_R. This is due to the fact that unlike fuzzy to binary, fuzzy to fuzzy considers more channels as available, using the threshold. Again P_R has a sharp rise, since fuzzy to fuzzy strategy assigns channels faster. The fuzzy to fuzzy strategy takes decision better and hence can handle congestion of higher level.

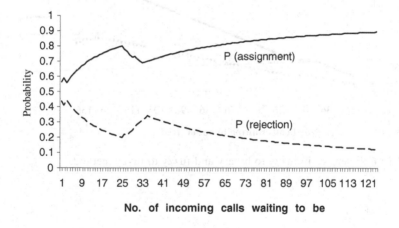

Fig. 3.12 The call drop probability and call assign probability

3.6.3 Comparison between the Methods

Fig. 3.13 shows a comparison between the fuzzy to binary and fuzzy to fuzzy methods in terms of call service. It is observed from this figure that no. of calls assigned (N_a) in fuzzy to fuzzy is always higher than that of fuzzy to binary scheme irrespective of the no. of waiting calls (N_s). Further, the slope of the fuzzy to fuzzy curve is also higher than that of fuzzy to binary.

Lastly, the saturation in the fuzzy to fuzzy system starts at N_s=134, while that in fuzzy to binary takes place at N_s= 120. Finally, the fall-off slope in fuzzy to binary scheme is higher than the fuzzy to fuzzy scheme, indicating that with a little increase in N_s the N_a will have a drastic fall-off in fuzzy to binary mapping scheme in comparison to that in fuzzy to fuzzy mapping scheme.

3.6.4   Call Rejection Threshold for Fuzzy to Fuzzy Module

The experimental shape of the curve for call rejection probability in fuzzy to fuzzy scheme is already explained in Fig. 3.12. In Fig. 3.14, we study the effect of variation of threshold in the fuzzy production rules deciding SHO. An apparent look into Fig. 3.14 reveals that the curves with threshold =0.75 and 0.5 are more or less parallel.

The increase in call rejection probability for increased threshold makes sense following rules1 to 6. The phenomena for threshold =0.25 however is apparently counter-intuitive as it lies between the two other curves, one with less, and the other with more threshold.

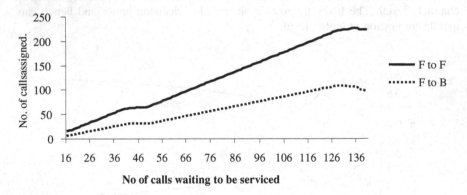

Fig.3.13 Call service in fuzzy to binary and fuzzy to fuzzy methods

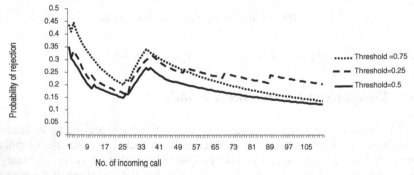

Fig. 3.14 Call rejection probability in fuzzy to fuzzy system with thresholds (0.5, 0.25, 0.75)

A detailed look into the call assignment strategy envisages that for threshold= 0.25 the feasibility condition, i.e. soft constraints, for channel assignment failed to be satisfied, giving rise to this phenomenon. A threshold of 0.5 seems to be an optimum choice as manages to satisfy the feasibility condition in one end and reduces the call rejection probability in the other end.

To compare the performance of our system with classical FCA [50] and DCA, we consider the call rejection probability of these two schemes in [2]. Reconstructing the same, and merging them with our results, we obtain Fig. 3.15. It is clear from Fig. 3.15 that both DCA and FCA have a high call rejection probability when no. of calls waiting for serviced (N_s) is very low. But it may be noted that the same cellular configuration can manage to handle almost double waiting calls, even with a significantly low call rejection probability. Thus our proposed scheme shows far better ability in reducing call block.

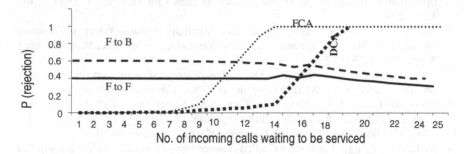

Fig. 3.15 Comparison of call drop probability in channel assignment problem

3.7 Conclusions

The proposed scheme is unique in its approach in combining Channel assignment with Call admission control, which is considered as two different areas of study. By implementing such concept we are able to minimize the interference in terms of co-channel, adjacency channel and co-site constraints, and could be able to ensure quality of service (QoS). Also the scheme implements reuse of channel strategy combined with forceful dropping of long calls, which improves the condition of a heavy traffic network and helps in avoiding congestion considerably. Not only the static calls, but the calls where the MS's are in a move, are also considered and are provided with proper service to minimize the drop of an ongoing call while moving. It has also taken care of the changeable conditions, including availability, feasibility, and hotness of a dynamic network. With all these aspects it gives a better result in managing calls than most of the existing systems.

References

1. Wong, D., Lim, T.J.: Soft handoffs in CDMA mobile systems. IEEE Personal Communications Magazine 4(6), 6–17 (1997)
2. Ye, J., Shen, X., Mark, J.W.: Call Admission Control in Wideband CDMA Cellular Networks by Using Fuzzy Logic. IEEE Transactions on Mobile Computing 4(2) (March/April 2005)
3. Kumar, D., Chellappan, C.: Adaptive Call Admission Control in Tdd-CDMA Cellular Wireless Networks. UbiCC Journal 4(3) (August 2009)
4. Jia, H., Zhang, H., Zhu, S., Xu, W.: Seamless QoS-Aware Call Admission Control Policy in Heterogeneous Wireless Networks. In: 4th International Conference on Wireless Communications, Networking and Mobile Computing 2008, vol. (12), pp. 1–6 (October 2008)
5. Huang, C.-J., Chuang, Y.-T., Yang, D.-X.: Implementation of call admission control scheme in next generation mobile communication networks using particle swarm optimization and fuzzy logic systems. Expert Systems with Applications 35(3), 1246–1251 (2008)
6. Ni, W., Li, W., Alam, M.: Optimal Call Admission Control Policy In Wireless Networks. In: Wireless Communications and Networking Conference, March 31-April 3, pp. 2975–2979 (2008)
7. Chen, Y.H., Chang, C.J., Shen, S.: Outage-based fuzzy call admission controller with multi-user detection for WCDMA systems. IEE Proc. Commun. 152(5) (October 2005)
8. Chung-Ju, Chang, L.-C., Kuo, Y.-S., Chen, S.S.: Neural Fuzzy Call Admission and Rate Controller for WCDMA Cellular Systems Providing Multirate Services. In: IWCMC 2006, Vancouver, British Columbia, Canada, July 3–6 (2006)
9. Malarkkan, S., Ravichandran, V.C.: Performance analysis of call admission control in WCDMA System with Adaptive Multi Class Traffic based on Fuzzy Logic. IJCSNS International Journal of Computer Science and Network Security 6(11) (November 2006)
10. Chen, H., Cheng, C.C., Yeh, H.H.: Guard-Channel-Based Incremental and Dynamic Optimization on Call Admission Control for Next-Generation QoS-Aware Heterogeneous Systems. IEEE Trans. Veh. Technology 57(5) (September 2008)
11. Liu, Z., Zarki, M.E.: SIR-Based Call Admission Control for DSCDMA Cellular System. IEEE J. Selected Areas of Comm. 12, 638–644 (1994)
12. Evans, J.S., Everitt, D.: Effective Bandwidth-Based Admission Control for Multiservice CDMA Cellular Networks. IEEE Trans. Vehicular Technology 48, 36–46 (1999)
13. Chang, C., Shen, S., Lin, J., Ren, F.: Intelligent Call Admission Control for Differentiated QoS Provisioning in Wide Band CDMA Cellular System. In: Proc. IEEE Vehicular Technology Conf. (VTC 2000), pp. 1057–1063 (2000)
14. Comaniciu, C., Mandayam, N., Famolari, D., Agrawal, P.: QoS Guarantees for Third Generation (3G) CDMA Systems Via Admission and Flow Control. In: Proc. IEEE Vehicular Technology Conf. (VTC 2000), pp. 249–256 (2000)
15. Bambos, N., Chen, S.C., Pottie, G.J.: Channel Access Algorithms with Active Link Protection for Wireless Communication Networks with Power Control. IEEE/ACM Trans. Networking 8, 583–597 (2000)
16. Andersin, M., Rosberg, Z., Zander, J.: Soft and Safe Admission Control in Cellular Networks. IEEE/ACM Trans. Networking 5, 255–265 (1997)

17. Ho, C.-J., Copeland, J.A., Lea, C.-T., Stüber, G.L.: On Call Admission Control in DS/CDMA Cellular Networks. IEEE Trans. Vehicular Technology 50, 1328–1343 (2001)
18. Fang, Y., Zhang, Y.: Call Admission Control Schemes and Performance Analysis in Wireless Mobile Networks. IEEE Trans. Vehicular Technology 51, 371–382 (2002)
19. Wu, S., Wong, K.Y.M., Li, B.: A Dynamic Call Admission Policy with Precision QoS Guarantee Using Stochastic Control for Mobile Wireless Networks. IEEE/ACM Trans. Networking 10, 257–271 (2002)
20. Fang, Y., Chlamtac, I.: Teletraffic Analysis and Mobility Modelling of PCS Networks. IEEE Trans. Comm. 47, 1062–1072 (1999)
21. Naghshineh, M., Schwartz, M.: Distributed Call Admission Control in Mobile/Wireless Networks. IEEE J. Select. Areas Comm. 14, 711–717 (1996)
22. Dziong, Z., Jia, M., Mermelstein, P.: Adaptive Traffic Admission for Integrated Services in CDMA Wireless-Access Networks. IEEE J. Seleced Areas of Comm. 14, 1737–1747 (1996)
23. Klir, G.J., Yuan, B.: Fuzzy Sets and Fuzzy Logic: Theory and Applications. Prentice-Hall (1995)
24. Wang, L.X.: A Course in Fuzzy Systems and Control. Prentice Hall (1997)
25. Rappaport, T.S.: Wireless Communications: Principles and Practice. Prentice Hall (1996)
26. Viterbi, A.J., Viterbi, A.M., Gilhousen, K.S., Zchavi, E.: Soft Handoff Extends CDMA Cell Coverage and Increase Reverse Link Capacity. IEEE J. Selected Areas of Comm. 12, 1281–1288 (1994)
27. Braae, M., Rutherford, D.A.: Fuzzy Relations in a Control Setting. Kybernetes 7, 185–188 (1978)
28. Del Re, E., Frantacci, R., Giambene, G.: Handover and Dynamic Channel Allocation Techniques in Mobile Cellular Networks. IEEE Trans. Vehicular Technology 44, 229–237 (1995)
29. Liu, T., Bahl, P., Chlamtac, I.: Mobility Modelling, Location Tracking, and Trajectory Prediction in Wireless ATM Networks. IEEE J. Selected Areas Comm. 16, 922–936 (1998)
30. Guerin, R.A.: Channel Occupancy Time Distribution in a Cellular Radio System. IEEE Trans. Vehicular Technology 36, 89–99 (1987)
31. Seskar, I., Maric, S., Holtzman, J., Wasserman, J.: Rate of Location Area Updates in Cellular Systems. In: Proc. IEEE Vehicular Technology Conc. (VTC 1992), pp. 694–697 (1992)
32. Nanda, S.: Teletraffic Models for Urban and Suburban Microcells: Cell Sizes and Handoff Rates. IEEE Trans. Vehicular Technology 42, 673–682 (1993)
33. Varshney, U., Jain, R.: Issues in emerging 4G wireless networks. IEEE Computer 34(6), 94–96 (2001)
34. Wong, D., Lim, T.J.: Soft handoffs in CDMA mobile systems. IEEE Personal Communications Magazine 4(6), 6–17 (1997)
35. Prakash, R., Veeravalli, V.V.: Locally optimal soft handoff algorithms. IEEE Transactions on Vehicular Technology 52(2), 231–260 (2003)
36. Lin, Y.-B., Pang, A.-C.: Comparing soft and hard handoffs. IEEE Transactions on Vehicular Technology 49(3), 792–798 (2000)
37. Hong, D., Rappaport, S.S.: Traffic model and performance analysis for cellular mobile radio telephone systems with prioritized and nonprioritized handoff procedures. IEEE Transactions on Vehicular Technology 35(3), 77–92 (1986)

38. Fang, Y., Chlamtac, I., Lin, Y.-B.: Channel occupancy times and handoff rate for mobile computing and PCS networks. IEEE Transactions on Computers 47(6), 679–692 (1998)
39. Orlik, P.V., Rappaport, S.S.: A model for teletraffic performance and channel holding time characterization in wireless cellular communication with general session and dwell time distributions. IEEE Journal on Selected Areas in Communications 16(5), 788–803 (1998)
40. Fang, Y., Chlamtac, I., Lin, Y.-B.: Call performance for a PCS network. IEEE Journal on Selected Areas in Communications 15(8), 1568–1581 (1997)
41. Jedrzycki, C., Leung, V.C.M.: Probability distribution of channel holding time in cellular telephone systems. In: Proc. IEEE VTC 1996, Atlanta, GA, vol. 1, pp. 247–251 (May 1996)
42. Zonoozi, M.M., Dassanayake, P.: User mobility modeling and characterization of mobility patterns. IEEE Journal on Selected Areas in Communications 15(7), 1239–1252 (1997)
43. Guerin, R.: Channel occupancy time distribution in a cellular radio system. IEEE Transactions on Vehicular Technology 35(3), 89–99 (1987)
44. Papoulis, A.: Probability, Random Variables, and Stochastic Processes. McGraw-Hill (1965)
45. Gross, D., Harris, C.M.: Fundamentals of Queueing Theory, 3rd edn. John Wiley & Sons, Inc. (1998)
46. Kelly, F.P.: Fixed point models of loss networks. Australian Mathematical Society 31, 204–218 (1989)
47. Vidyarthi, G., Ngom, A., Stojmenovic, I.: A Hybrid Channel Assignment Approach Using an Efficient Evolutionary Strategy in Wireless Mobile Networks. IEEE Transactions on Vehicular Technology 54(5), 1887–1895 (2005)
48. Battiti, R., Bertossi, A.A., Brunato, M.: Cellular Channel Assignment: a New Localized and Distributed Strategy. Mobile Networks and Applications 6, 493–500 (2001)
49. Li, S., Wang, L.: Channel Assignment for Mobile Communications Using Stochastic Chaotic Simulated Annealing. In: Mira, J., Prieto, A.G. (eds.) IWANN 2001. LNCS, vol. 2084, pp. 757–764. Springer, Heidelberg (2001)
50. Sivarajan, K.N., McEliece, R.J., Ketchum, J.W.: Channel assignment in cellular radio. In: Proc. 39th IEEE Veh. Technol. Soc. Conf., pp. 846–850 (May 1989)
51. Sivarajan, K.N., McEliece, R.J., Ketchum: Dynamic channel assignment in cellular radio. In: Proc. 40th IEEE Vehicular Technology Conf., pp. 631–637 (1990)
52. Ngo, C.Y., Li, V.O.K.: Fixed channel assignment in cellular radio networks using a modified genetic algorithm. IEEE Trans. Veh. Technol. 47(1), 163–172 (1998)
53. Smith, K., Palaniswami, M.: Static and dynamic channel assignment using neural network. IEEE Journal on Selected Areas in Communications 15(2) (February 1997)
54. Pedryz, W., Vasilakos, A.: Computational Intelligence in Telecommunication Networks. CRC Press, Boca Raton (1997)
55. Pedrycz, W., Gomide, F.: An Introduction to Fuzzy Sets: Analysis and Design. MIT Press (1998)

Chapter 4
An Evolutionary Approach to Velocity and Traffic Sensitive Call Admission Control

The chapter proposes a new approach to call admission control in a mobile cellular network using an evolutionary algorithm. Existing algorithms on call admission control either ignore both variation in traffic conditions or velocity of mobile devices, or at most consider one of them. This chapter overcomes the above problems jointly by formulating call admission control as a constrained optimization problem, where the primary objective is to minimize the call drop under dynamic condition of the mobile stations, satisfying the constraints to maximize the channel assignment and minimize the dynamic traffic load in the network. The constrained objective function has been minimized using an evolutionary algorithm. Experimental results and computer simulations envisage that the proposed algorithm outperforms most of the existing approaches on call admission control, considering either of the two issues addressed above.

4.1 Introduction

Call Admission Control (CAC) refers to an interesting decision-making problem of efficient call management in a mobile cellular network. The primary objective of this problem is to serve as many calls as possible, and prevent dropping of calls in progress [1-3]. Additionally, an efficient call management system also aims to assign appropriate channels to the incoming/handoff calls, so that the necessary soft constraints for channel assignment are maintained [1-8].

Typically, soft constraints include co-channel, co-site, and adjacent channel constraints, all of which need to be satisfied to serve the secondary objective. In the current literature, Quality of Service (QoS) is often used to measure the quality of CAC with an attempt to maximize call assignment and soft handoff, satisfying the soft constraints. The higher the QoS, the better the CAC.

This chapter provides a novel approach to formulate the CAC as a complex decision-making problem with an objective to optimize service of calls in a dynamic environment under the fluctuation of load and motion of the mobile station. The formulation involves construction of an objective function with constraints, and has been solved using an evolutionary algorithm. Experiments have extensively been undertaken to minimize call drops.

S. Ghosh and A. Konar: CAC in a Mobile Cellular Network, SCI 437, pp. 129–152.
springerlink.com   © Springer-Verlag Berlin Heidelberg 2013

4.1.1 Review

Limited works on CAC employing genetic algorithm (GA) has appeared in the existing literature. One of the initial works utilises a local policy, where the cells are square (and not hexagonal) in shape and the call arrival rate follows Poisson distribution in a Markov model [1]. The performance of the system depends only on the linear combination of call dropping and the call blocking probability. The policy is said to be local since a base station only gets the information of its four neighbours. Here a two parent two offspring GA is used. Clique packing technique is used to take care of the channel assignment problem implicitly [1]. The process is terminated when the policy improvement appears to stagnate.

In the year 2000 [7, 8] the authors proposed a scheme, where the system is defined in terms of m classes of users in each cell. A user in class i would require b_i amount of bandwidth. The goal was to maximize $\sum x_i b_i$ for candidate solutions x_i, where the handoff and blocking probability have a given upper bound. Both semi Markov decision process and GA are used. Here also the process is local one considering the call acceptance of a single cell.

In 2004 one more CAC scheme was proposed for General packet radio service (*GPRS*) using GA [6]. Here the GPRS architecture was of given special emphasis. The CAC module is almost the same as the one described in [7]. But the fitness function is different as it is the product of QoS factor and length of the queue of the class and inverse of the square root of frequency. The advantage of this method over the previous one described in is the usage of GPRS technology and its architecture [6].

4.1.2 The Problem

CAC systems are mainly used to take decisions whether a call should be serviced, blocked or dropped by a Base Station (BS), and if serviced, it identifies the channel to be assigned to that call. At the time of taking such decision, the interference is considered the only factor in most of the current literature [1 - 48].

Existing literatures presented above, concentrates on specific aspects of the call admission control problem. A few of these considers rectangular cells and the near neighbour and hence ignores the entirety of the network, while the rest considers only the hand off and call drop due to insufficient channels. Unfortunately, most of the above literatures consider the mobile stations either static or moving very slowly in the small area with very low traffic load. The mobility factor comprising speed and direction of movement has been ignored in these works. The overall network load is also ignored and hence the handoff policies used have a partial effect on the network. Most of the above schemes are localized to a single cell. Hence the channel reuse is also not done efficiently.

In this chapter, we propose a scheme that takes a more wide view of the call admission control problem. We consider our cells to be hexagonal so as to easily track the movement of Mobile Stations (MS) in the neighbourhood cells. Instead of considering a single cell scenario we have taken a small network to implement the algorithm to incorporate the intercellular communication efficiently. The decision of acceptance or rejection of a call dose not only depend on the feasibility and availability of the channel but also on the speed at which the MS moves and its direction of movement. Its geographical location with respect to the base station also has a significant importance. Moreover the traffic density on a cell is also considered to be a determining factor.

The reuses of the channels also make the approach more effective. The algorithm proposed in the chapter is based on all the factors mentioned above which more perfectly handles the real world situation.

There are certain parameters to measure the efficiency of such a system. One important parameter is the number of call drop that shows the number of unwanted call disconnection in the system and should be as low as possible. The other one would be the call block where the new call attempts are rejected by the network due to the insufficiency of the free channels. The number of assignable channels that helps in increasing the efficiency of the system is another important parameter considered here.

In this chapter, we propose a scheme, which takes into account the motion aspect of the mobile station as well, besides considering all the necessary objectives of CAC. We here use three parameters: i) speed, ii) direction, and iii) distance of the MS from the nearest base station to model the motion of the mobile station. The above three parameters play an important role at the time of soft handoff of a call from one cell to other. The importance can be explained with the help of Fig. 4.1, where the central cell has six neighbours.

The MS in such a cell while in service may move in various directions with different speed. If it moves toward cell 5 or 6 directly then the channel for handoff will be searched in those cells. When it moves slowly along the common boundary of cell 5 and 6, the cell with base station nearer to the current location of MS is considered. Again, if the above movement takes place with a very high speed, then the call may be dropped due to high interference. Suppose it moves towards cell 2 very slowly, then there may not be a requirement for a soft handoff at all since it may never cross the existing cell boundary.

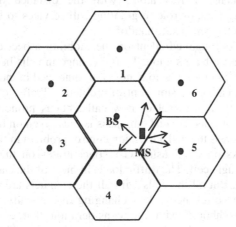

Fig. 4.1 Illustrating the need to consider speed in a given mobile cellular network.

Hence, we aim at developing an optimal set of assignment, which will address of the quality of service, and the velocity aspect of the scheme.

4.1.3 The Approach

The problem of call admission can be reconstructed as an optimization problem with given constraints. As such a fuzzy approach would not be appropriate in that case. Graph colouring problem and others are not suitable since there are some resources that are shared by all the cells and hence using a unique colour to represent a cell is difficult. Hence the natural option was to take a GA approach.

Assigning a call to a channel is done using the electromagnetic compatibility constraints. In general, there are three types of constraints [3]:

- **Co-channel Constraint (CCC):** The same channel cannot be simultaneously allocated to a pair of cells unless there is a minimum geographical separation between them.
- **Adjacent Channel Constraint (ACC):** Adjacent channels cannot be assigned to a pair of cells unless there is a minimum distance between them.
- **Co-site Constraint (CSC):** A pair of channels can be employed in the same cell only if there is a minimum separation in frequency between them.

These constrain are said to be the soft constrains of channel assignment and are used as the feasibility constrain for call admission control. Call admission control involves decision-making regarding assignment of channels to incoming calls or dropping calls in service, besides serving handoff calls, if feasible. To arrive at such decision, certain factors should be taken into consideration.

When a new call arrives, the feasibility condition for call assignment is checked using the soft constraints and the decision about call assignment/blocking is taken.

When a MS starts moving with a call in service the scenario changes. Then the feasibility condition only is not adequate to prevent call dropping. The speed and direction of movement of MS along with the distance of MS from the neighbouring BS play a major role in granting call services to soft-handoff since the crossover of cells triggers a soft-handoff.

The previous works remained silent on the mobility aspect of the MS and its location from the nearest base station. Traffic change in cells because of variation in incoming calls is a common phenomenon, but unnoticed in most of the literature while addressing the call admission control problem. Traffic change, however is not only influenced by the arrival of new calls, but is pronounced also by the mobility of the existing calls. Special emphasis is thus given on traffic changes and mobility aspect to address the call admission control problem in this chapter.

Most of the works on CAC using GA concentrates on the decision-making about the calls in a single cell. The traffic loads in most of the cases are considered to be static in nature. But all the cells with all the channels are going through the same process and hence the network is changing dynamically. Considering only one cell and its neighbours with the consideration that conditions of those neighbours are static also makes the system limited in effect. If the whole network is considered, then the solution becomes more applicable and robust.

In this chapter we have considered the case of dynamically changing cells of the entire network. We have considered dynamically changing demand of the channels as well as random movement of each MS with varying speed. In this context we see that the call drop due is minimized considerably and soft handoff is achieved effectively. The call service is almost consistent with the change in demand.

The proposed system thus is more sensitive to the continuously changing demands of the real world and the changing location of the MS. With this the QoS is also taken care of using the soft constraints for interference. This approach is a holistic one as all the cells of the network is considered. Hence it is a new way of looking into the CAC problem, which caters better in a real world situation.

4.2 Formulation

Here we have attempted to handle the calls with their mobility factor. A caller may be in motion when a call is generated or while being serviced. Depending on the speed of the MS and the direction in which they are moving, there may be a need to reassign the call to a different channel of a different cell. To explain such situation we start with defining the primitives.

4.2.1 Definitions

We consider a system of M hexagonal cells present in the network and each of them has N number of frequency channels. In CAC we need to find out the best allocation of calls in different cells. So we need to know about the current status of allocation, which is usually represented by an allocation matrix. In this chapter, we plan to select the appropriate allocation matrix online, so as to satisfy given objective function and systems constraints to be introduced later. The formal definition of allocation matrix is given here for convenience.

Definition 1: Let $[f_{m,\,i}]$, $\forall i \in M, m \in N$, be a binary matrix describing allocation of channels in given cells, where

$$F = [f_{m,i}] = \begin{cases} 1, & \text{if } m^{th} \text{ channel is allocated to serve a call in the } i^{th} \text{ cell} \\ 0, & \text{if the } m^{th} \text{ channel in the } i^{th} \text{ cell is free.} \end{cases}$$

Example 4.2.1: An example of a 4x3 allocation matrix where there are 4 cells and 3 channels is given as

$$\text{channel} \quad \rightarrow$$

$$\text{Cell} \quad \downarrow \quad \begin{bmatrix} 0 & 1 & 0 \\ 0 & 0 & 0 \\ 1 & 0 & 0 \\ 0 & 0 & 1 \end{bmatrix}$$

This implies that the 2nd channel of the 1st cell is serving a call and rest of the channels are free.

The knowledge of calls assigned to channels in any cell is a very important to measure the feasibility of assigning the new calls satisfying the soft constraints.

Definition 2: Let *assignment matrix* $A = [a_{m,i}]$, $\forall i \in M$, $m \in N$, be a matrix describing the assignment of calls to the channels of a cell, where $a_{m,i} = p$, *where p is the number of channel assigned to the* m^{th} *call in cell i*.

Example 4.2.2: Let there be 10 calls in the network with 4 cells and 3 channels. Then

$$
\text{calls} \rightarrow
$$

$$
\text{Cell} \downarrow
\begin{bmatrix}
1 & 0 & 0 & 0 & 2 & 0 & 0 & 0 & 0 & 3 \\
0 & 0 & 1 & 0 & 0 & 0 & 0 & 0 & 2 & 0 \\
0 & 2 & 0 & 0 & 0 & 1 & 0 & 3 & 0 & 0 \\
0 & 0 & 0 & 1 & 0 & 0 & 3 & 0 & 0 & 0
\end{bmatrix}
$$

We can see that the 1st call is served by the 1st channel of the 1st cell, the 2nd call by the 2nd channel of 3rd cell, 3rd call by the 1st channel of 2nd cell and so on.

While allocating the calls to the channels we should maintain the QoS in terms soft constraints. The soft constrains ensure that the new call assignment do not have any interference with the existing calls assignment in the neighbouring cells or in the same cell. In this chapter, we measure the QoS as a function of three important network attributes: feasibility, hotness (the measure of load) and motion of the MS. The feasibility of channel assignment is often expressed as linear combination of allocation and compatibility matrices. A formal definition of compatibility matrix is given here.

Definition 3: The *compatibility matrix* C gives a measure of satisfaction of the soft constraints, attempted to minimize co-channel, co-site and the adjacency-channel interference, whose non-diagonal and diagonal elements are expressed by

$C_{j,k}$ = *minimum channel seperation required for a call assignment in cell j when*

     there are calls assigned in cell k, and

$C_{i,i}$ = min *imum channel seperation required to assign a call in*

     cell i when there are other channels assigned in the same cell.

Speed of the MS is a very important aspect of the CAC since it affects the soft handoff and the call drop process.

Example 4.2.3: A compatibility matrix in a four cell network is given as

$$\begin{bmatrix} 4 & 3 & 2 & 0 \\ 3 & 4 & 3 & 2 \\ 2 & 3 & 4 & 3 \\ 0 & 2 & 3 & 4 \end{bmatrix} \quad \begin{array}{l} where,\ C_{i,i}\ =4 \\ and \quad C_{i,j}\ \in \{0\ to\ 3\} \end{array}$$

Definition 4: Speed $V = [v_{p,i}]$ in the present context refer to rate of position changing of a MS busy with a call utilizing a channel p in cell i.

Example 3.2.4:

$$v_{p,i} = 0 \qquad \qquad when\ the\ MS\ is\ static$$

$$0 < v_{p,i} \le 60 \qquad when\ inside\ the\ city$$

$$v_{p,i} > 60 \qquad \qquad in\ highways$$

Distance of a MS from the BS is also an important aspect as it takes part in the soft handoff process.

Definition 5: Distance of MS j from the BS i $\forall i \in M$ is denoted as $Dis = [dist_{i,j}]$

Example 3.2.5: $dist_{i,j} = d$ (Fig. 4.2)

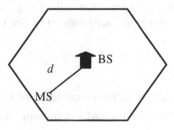

Fig. 4.2 Measurement of d

Hotness is another major factor while deciding about a call assignment to a channel. It is named hotness since it shows how many calls are coming to a particulate cell i.e. how hot and happening the cell is. If the number of incoming calls becomes very high the limited resource of the system will not be able to handle the incoming calls and thus subsequent new calls will be blocked.

Definition 6: Hotness of a cell i is defined as the number of incoming calls per unit time and is denoted as $H = [h_i]$

Example 3.2.6: Hotness in a 4-cell scenario where maximum number of incoming calls is 10 can be given as

$$H = [h_i] = \begin{bmatrix} 2 & 10 & 8 & 5 \end{bmatrix}$$

The angle of motion of MS with respect to the BS shows the direction in which the MS is moving and hence the search for cells with free channels becomes easier. Hence it is also an important factor in CAC.

Definition 7: Angle of motion is the angle made by the direction of motion of a MS p with respect to the BS i and is denoted by $Delt = [\delta_{p,i}]$, and is illustrated in Fig.4.3.

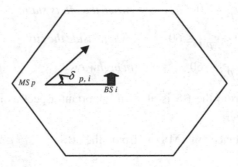

Fig. 4.3 Angle of Motion

Call duration is useful for finding out the calls going for a very long time. At the time of high congestion, these calls are dropped to free the channels for reuse.

4.2.2 Formulation

Let the time taken by each call in a given channel p, of each cell i is denoted by $T_{p,i}$. If $v_{p,i}$ be the speed with which an MS is moving in any direction, then, its velocity along that direction is $v_{p,i} \cos \delta_{p,i}$. Again if $dist_{p,j}$ be the distance traversed by the MS in time $\Delta T_{p,i}$ then average probability of capturing that MS is denoted as P_{CMS} and is given by

$$P_{CMS} = \sum_p \frac{(v_{p,i} \cos \delta_{p,i}) \Delta T_{p,i}}{dist_{p,j}} \tag{4.1}$$

An increase in the value of the above expression is created by a high range of velocity, which gives very little time to search a new channel and to go for a soft hand off (SHO). Hence it should be minimized.

The difference between the two calls in the two different channels should be at a minimum distance to avoid the interference described as soft constrains above. The feasibility of assignment of calls in a cell i can be checked by satisfying inequality (4.2a) [1], [3]. In (4.2a) the left hand side of the inequality suggest the distance between two channels of which one assign m^{th} call in i^{th} cell and the other assign n^{th} call in j^{th} cell. The right hand side of the inequality gives the minimum channel separation required to assign calls in both the cells i and j. The condition demands that the channel separation should satisfy the bare minimum value obtained from compatibility matrix and is said to be feasibility condition and denoted by Feas.

$$\sum_m \sum_n (\sum_i (|a_{m,i} - a_{n,j}|)) < \sum_i C_{i,j} \qquad (4.2a)$$

$$\Rightarrow \text{Feas}^j = \sum_m \sum_n (\sum_i (|a_{m,i} - a_{n,j}| - C_{i,j})) < 0 \qquad (4.2b)$$

It follows from the last inequality that smaller the value of the left hand side, lesser is the interference and thus better is the quality of service of the call assigned to the channel.

The traffic load in a cell is an important issue to determine the possible admission of incoming calls in a given cell. Traffic load in the cell i may be expressed as the ratio of incoming calls and the total free channels of the cell. We define a metric to measure the traffic load in a given cell i, denoted as Load and is given by

$$Load = \frac{h_i}{\sum_m (1 - f_{m,i})} \qquad (4.3)$$

Hence the traffic load is the more if the number of incoming calls exceeds considerably than the free cells in the system and starts affecting the overall system performance. Hence this is also to be minimized.

We now construct an objective function, the minimization of which yields a possible solution to the call admission problem. Since expressions (4.1), (4.2b) and (4.3) all need to be minimized, a minimization of their linear combination offers an objective that jointly satisfies all the three basic objectives. The overall objective function is given by

$$Z_{Candidate} = P_{CMS} + Feas + Load$$

$$= \sum_i \left(\sum_p \frac{(v_{p,i} \cos\delta_{p,i})\Delta T_{p,i}}{dist_{p,i}} + \sum_m \sum_n (\sum_j (|a_{m,i} - a_{n,j}| - C_{i,j})) + \frac{h_i}{\sum_m (1 - f_{m,i})} \right) \qquad (4.4)$$

The lower the value of the function better is the performance of CAC system. Here we also define the difference of fitness ΔZ as

$$\Delta Z = Z_{offspring} - Z_{parent} \qquad (4.5)$$

4.3   Proposed Algorithm

We have used the traditional concept of GA for the call admission control system. Here we have considered the call assignment in the network as the parent population and used crossover and mutation to generate the offspring population. Since it is considered on the total network the chromosomes are two dimensional in nature with rows representing cells and columns representing columns. With the generation of offspring we calculate the cost of both parents and offspring and retain the fittest.   The algorithm is described below with the abbreviations.

M, N	Total number of channel and cell;
ITE	Total number of iteration;
L	Long call interval;
ε	Very small number;
s	Size of population;
V	Speed as given in definition 4
H	Hotness as given in definition 6
A	Assignment matrix as given in definition 2
C	Compatibility matrix as given in definition 3
Delt	Angle of motion as given in definition 7
Dist	Distance as given in definition 5.
P	$\{p^j\}$ = Initial population.
C-Pool	Parent Pool.
O-Pool	Offspring Pool
Z	fitness for a candidate

Algorithm:

CALL_Adm(A,C;Z)
  BEGIN
FOR (j=0 ; j<s; j++)
   Initial (A, C; p^j) ;
  END FOR.
  P={p^j};
  FOR (i =0;i<ITE; i++)
   Do

     Selection (A, C, P; C-Pool);
    Crossover (C-Pool; O-Pool);
Mutation (O-Pool; O-Pool);
FOR (j=0; j<s; j++)
   Fit (A, C, Dist, Delt , V, H, C-Pool ; Z $_{Pool}$) ;
    Fit (A, C, Dist, Delt , V, H, O-Pool ; Z $_{Pool}$) ;
   IF ($\Delta Z_f^j > 0$)
Update C- Pool as $p^j = f^j$;
    END IF.
    END FOR.

$(T_{x,i,j})++;$

$$IF\left(T_{x,\ i,j} > L\right)$$

$\quad\quad$ **Drop_call** *(x, i, j ,O-Pool; O-Pool);*
$\quad$ *END IF;*
$\quad\quad\quad$ *IF $(\Delta Z_f^j < \varepsilon)$*
$\quad$ *BREAK;*
$\quad\quad$ *END IF;*
$\quad\quad\quad$ *WHILE (CONDITION 2a)*
END FOR;
$\quad$ *END.*

Initialization:

In this chapter, we consider the solutions as the assignment of calls to channels of all the cells in the network. Hence we have each solution as a *M x N* vector F (assignment) such that

$$F = \{f_{m,i}\} = \left\{ \begin{array}{l} 1 \ \ when \ m^{th} \ channel \ in \ i^{th} \ cell \ is \ in \ service \\ 0 \ otherwise \end{array} \right\},$$

where F satisfy the soft constraints.

$\quad$ We have considered a pool of *p* such vectors of assignment as a parent population all of which satisfies the soft constraints.

Initial *(A, C; p^j):*
$\quad\quad$ *BEGIN.*
$\quad$ *FOR $\left(i = 0; i < N; i++\right)$*
FOR $\left(m = 0; m < M; m++\right)$
$\quad$ *IF $\left(\| a_{m,i} - a_{n,j} \| < C_{i,j}\right)$*

$$f_{m,i} = \left\{ \begin{array}{l} 1 \ when \ m^{th} \ channel \ in \ i^{th} \ cell \ is \ in \ service \\ 0 \ otherwise \end{array} \right\},$$

$\quad\quad$ *END IF.*
$\quad\quad$ *END FOR.*
$\quad$ *END FOR.*
$\quad$ *Construct $p^j = F = \left[f_{m,i}\right]$*
$\quad$ *Return p^j.*
$\quad\quad$ *END.*

Example 3.3.1: We consider a network with 3 cells and 4 channels. Then the parent matrix, obeying the compatibility constraints, may look like the following p^j where a '1' represents a channel in a cell is assigned to a call and a '0' represents a free channel.

$$p^j = F = [f_{m,i}] = \begin{bmatrix} 0 & 1 & 0 \\ 0 & 0 & 0 \\ 0 & 0 & 1 \\ 0 & 0 & 0 \end{bmatrix} \quad \text{where, M=3 and N=4.}$$

Selection:

A roulette wheel with slots sized according to feasibility is used for selection process. We construct such a roulette wheel as follows

- First we calculate the feasibility value for a parent p^j using condition (2b) denoted as $feas^j$.

- Then we calculate the total feasibility of all parents $\sum_{j=0}^{s-1} feas^j$.

- Now we calculate the probability of selection of a parent p^j

as $\text{Pr} ob_{p^j} = \dfrac{feas^j}{\sum_{j=0}^{s-1} feas^j}$.

- Now we calculate the cumulative probability $CP_i = \sum_{j=0}^{i} \text{Pr} ob_{p^j}$.

- The selection process is based on spinning the roulette wheel s times; each time we select a single chromosome for a new population in the following way:

  o   Generate a random (floating) number r in the range [0; 1],
  o   If $r < CP_1$ then select the 1st chromosome (p^1);
  o   Otherwise select the ith chromosome p^i (2< i < s) if $PROB_{i-1} < r < PROB_i$.

This way we can choose the parents who have higher feasibility of getting new calls in the population. The pseudo code for selection is given below.

Selection (A, C, P; C-Pool):

Let x be an integer.
BEGIN.

$$Feas^j = \sum_m \sum_n (\sum_i (| a_{m,i} - a_{n,j} | - C_{i,j}))$$

   $Fsum = 0;$

```
FOR (j=0 ; j< s ; j++)
     Fsum = Fsum + Feas^j;
END FOR;
```

$$\Pr{ob}_{p^j} = \frac{feas^j}{\sum\limits_{j=0}^{s-1} feas^j}$$

$CP_i = 0.$

```
FOR ( j = 0 ; j< i ;  j++)
```

$$CP_i = \sum_{j=0}^{i} \Pr{ob}_{p^j}$$

```
   END FOR.
```

$r = Random (float in [0, 1]).$

$IF(r < CP_0)$ then $x=0;$

```
 ELSE

     FOR (i=1; i < s ; i ++)
```

$IF (PROB_{i-1} < r < PROB_i)$

```
       x= i;
                Pool = Pool U {p^x};
                BREAK;
         END IF.
         END FOR.
END IF.
 Return C-Pool;
END.
```

Crossover:

We use the C-Pool generated by the selection procedure for crossover. We consider two random chromosomes from the C-Pool to crossover. We consider a one-point crossover and generate a random position *(r, c)* at which the crossover takes place to produce two offspring.

Let $p_{m,l}^x$ and $p_{m,l}^y$ be two parents taking part in crossover. After crossover the offspring are kept in O-Pull.

The pseudo code of crossover is given as:

Crossover (C-Pool; O-Pool):

```
BEGIN.
   FOR(ps=0; ps < SizeOf( C-Pool) ; ps ++)
```

```
// Taking chromosomes for cross over
x = Random (number < SizeOf ( C-Pool));
y = Random (number < SizeOf ( C-Pool) && ≠x);
// Deciding position of cross over
r = Random (number < m);
c= Random (number < l);
FOR( i= 0; i < r; i++)
FOR (j =0; j < c; j++)
```

$$f^{\,x}_{\,j,i} \;=\; p^{\,x}_{\,j,i}$$

$$f^{\,y}_{\,j,i} \;=\; p^{\,y}_{\,j,i}$$

```
END FOR.
    END FOR.
FOR ( i= r ; i < m; i++)
        FOR (j =c; j < l; j++)
```

$$f^{\,x}_{\,j,i} \;=\; p^{\,y}_{\,j,i}$$

$$f^{\,y}_{\,j,i} \;=\; p^{\,x}_{\,j,i}$$

```
    END FOR.
    END FOR.
        Update O-Pool;
    END FOR.
END.
```

Example 3.3.2: Let the parents have 3 cells and 4 channels and the crossover point is randomly chosen as 3rd row 3rd column.

i.e r_{max} $= 4$ and c_{max} $= 3$ and

$$p^1 = \begin{matrix} 0 & \boxed{1 & 0} \\ 0 & \boxed{0 & 0} \\ 1 & 0 & 0 \\ 0 & 0 & 1 \end{matrix} \qquad\qquad p^2 = \begin{matrix} 1 & \boxed{0 & 0} \\ 0 & \boxed{0 & 1} \\ 0 & 0 & 0 \\ 0 & 1 & 0 \end{matrix}$$

Then after crossover at r = 3 and c =3 the offspring become

$$f^1 = \begin{matrix} 0 & 1 & 0 \\ 0 & 0 & 1 \\ 0 & 0 & 0 \\ 0 & 1 & 0 \end{matrix} \qquad \text{and} \qquad f^2 = \begin{matrix} 1 & 0 & 0 \\ 0 & 0 & 0 \\ 1 & 0 & 0 \\ 0 & 0 & 1 \end{matrix}$$

Mutation:

We consider the random call hang-up in the system and express this phenomenon as mutation. Suppose in a given cell r, a call served with channel c is disconnected by the caller. Then following the definition of allocation matrix F, we understand that the element $f_{ml} = 0$ for m= r and l=c, after disconnection of the call.

$$\text{i.e. } f_{m,l} = 0, \text{ where } m = r, l = c$$

We have considered minimum two such hang-ups in the network. Accordingly random positions are generated where mutation is done as described above.

Mutation (O-Pool; O-Pool):

BEGIN.
FOR (i=0; i < Size Of (O-Pool); i++)
        r = Random (number < m);
        c= Random (number < l);

$$f_{r,c} = 0,$$

   END FOR
   Update O-Pool;
END.

Example 3.3.3: Let there be mutation at the positions (1,2) and (4,3) for the parents with 4 cells and 3 channels
   i.e, r_{max} = 4 and c_{max} = 3 and

$$f = \begin{matrix} 0 & 1 & 0 \\ 0 & 0 & 0 \\ 1 & 0 & 0 \\ 0 & 0 & 1 \end{matrix}$$

Now, r1= 1, c1=2, and r2 = 4, c2 = 3. Hence after mutation the offspring will become

$$f = \begin{matrix} 0 & \boxed{0} & 0 \\ 0 & 0 & 0 \\ 1 & 0 & 0 \\ 0 & 0 & \boxed{0} \end{matrix}$$

Fitness check :

This is the part where we calculate the fitness of the new offspring and there parents using the equation (4) where all the other values are known.

Fit (A, C, Dist, Delt , V, H, Pool,; Z $_{pool}$) :
BEGIN.
WHILE(! EOF(O-Pool)

$FOR\ (p=0\ ;\ p <N;\ p++)$

$$P^i_{CMS} = \sum_p \frac{(v_{p,i}\cos \delta_{p,i})\Delta T_{p,i}}{dist_{p,i}}.$$

   $END\ FOR.$

  $FOR\ (m=0\ ;\ m <N;\ m++)$

$FOR\ (n=0\ ;\ n <N\ \&\&\ m \ne n\ ;\ n++)$

  $FOR\ (j=0\ ;\ j <M;\ j++)$

$$Feas^i = \sum_m \sum_n (\sum_j (|a_{m,i} - a_{n,j}| - C_{i,j}))$$

 $END\ FOR$

$END\ FOR.$

    $END\ FOR.$

    $FOR\ (m=0\ ;\ m <N;\ m++)$

$$Load^i\ \frac{h_i}{\sum_m (1 - f_{m,i})}$$

   $END\ FOR.$

$$Z_{Pool} = \sum_i P^i_{CMS} + Feas^i + Load^i$$

$RETURN(Z_{Pool}).$

  $END\ WHILE$

$END.$

Call Drop: this is where the call is forcefully terminated due to the time constraint.

Drop_call $(x, i, O\text{-}Pool;\ O\text{-}Pool)$

 $BEGIN$

$$f^x_{i,\,j} = 0;$$

 END

4.4   Experiments and Simulation

In this experiment we have considered the following assumptions:

- The network has 21 hexagonal cells and 7 channels(Fig 4.4)
- The value of
  - o Co-channel distance is 2
  - o Adjacency channel distance is 3
  - o Co-site distance is
- The number of incoming calls lays in the range 0 to 150 and changes dynamically.

- Initial population size was taken as 20.
- The velocity change was from 0 to 120 km/hr.
- The distance between two base stations is 2 km and remains unchanged.
- The calls are considered long if they go on more than 30 minutes.
- The direction of the MS, its distance from the base station and velocity changes dynamically.

The algorithm stated above is executed using the given fitness function. The initial conditions and necessary changes of the dynamic network are enforced obeying the above stated assumptions. The results obtained are shown an explained below. Fig.4.5 depicts the fitness values with respect to the increasing demand. In the initial stage when the demand is low the fitness starts with a low value and further goes down. Hence more and more channel start getting assigned.

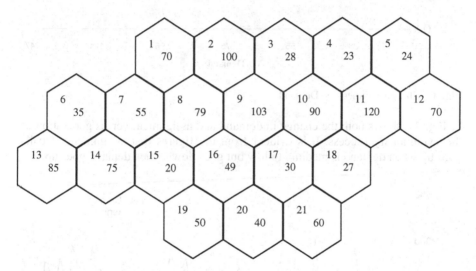

Fig. 4.4 Network of 21 cells

Initially the fitness starts decreasing as channels starts getting assigned to the calls. The fitness goes down after shooting up to a highest peak. This is due to a strong congestion in the system due to the insufficiency of the free channels in the network. This situation is resolved by dropping calls which are in service for long intervals. Again the fitness goes down and calls get assigned to channels. The in-between fluctuations in fitness values are due to the change in velocity as well as the distance of MS from BS. But as traffic load increases, the system is not able to handle the situation due to in sufficient free channels.

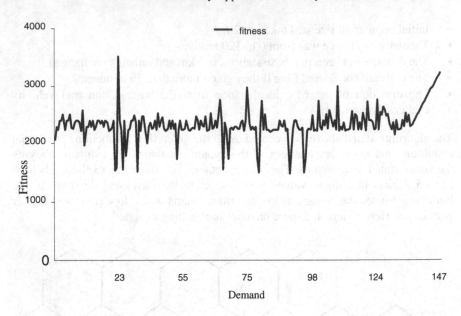

Fig. 4.5 Changing Fitness vs. Demand

Fig. 4.6 shows both the change in demand and assignment versus generation of the evolutionary process. It is evident from the figures that assignment of calls goes up when demand is medium to low but goes down when demand goes up.

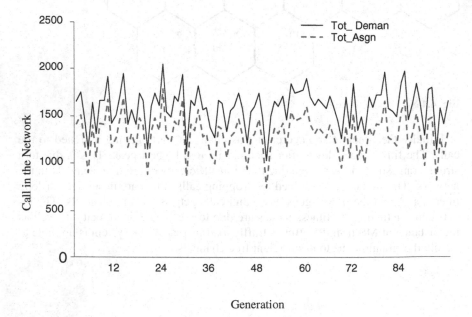

Fig. 4.6 Call Assignment with changing demand over generations

Again after a bulk of call assignment the number of available channels decrease and then new calls get blocked. This happens due to the high level of congestion and insufficient channels. After a while some calls end naturally and the channels are again free. So call assignment rate again increases.

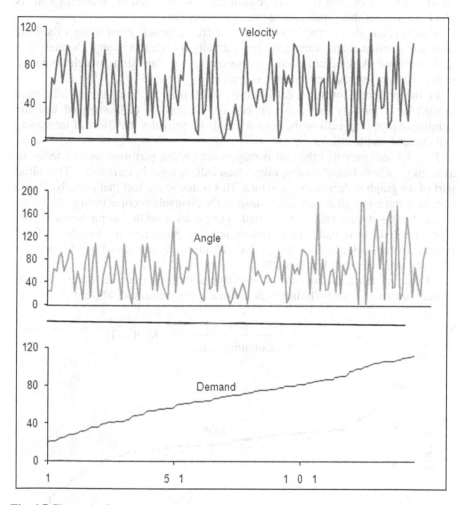

Fig. 4.7 Change in fitness due to change in the parameters

Fig. 4.7 shows the change in different parameters of the dynamic network with respect to increasing demand. We have here a randomly changing velocity and directions of calls in each channel. We calculate the fitness of assignment in such changing system with an increasing demand.

It is seen that whenever the velocity as well as the angle changes sharply the fitness function increases sharply even though the demand is not very high. Again when the angle changes sharply again there is a rise of fitness value. Again it is evident from Fig. 4.7 that fitness value goes down for slower speed and smaller change in direction. But if the congestion level is very high i.e. incoming calls is plenty in number then fitness value rises and stays flat.

This happens due to the insufficiency of free channels even though the long calls are terminated and some calls end naturally by releasing some channels free in the network. A flood of in coming calls can create congestion, which cannot be handled as long as resources are not increased.

In the present context of call admission control, the measure of efficiency (*MOE*) is defined as the ratio between the number of calls serviced and the number of incoming calls in the network. Fig. 4.8 provides the effect of increasing call demand on the efficiency.

Fig. 4.8 indicates that the call management system performs well in terms of assigning calls to free channels, even when call demand is increasing. The initial part of the graph is decreasing in nature. This is due to the fact that initially, after a certain number of call assignments, most of the channels become occupied.

At that point new calls are rejected. Then after a while, as proposed by our algorithm long time calls are retrieved and this procedure recurs after a fixed interval of time. Hence after this period call drop are minimized. And hence the MOE becomes more of less smooth.

At the end when the number of incoming calls increases excessively that the system cannot cope up with the given resources, the measure of MOE drops down.

$$MOE = \frac{\text{No. of call serviced}}{\text{No. of incomming calls}} \quad (0 \le MOE \le 1)$$

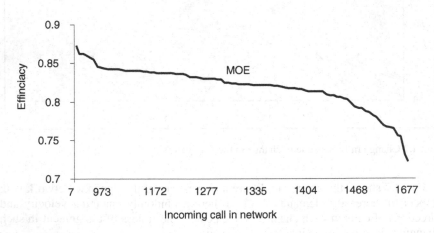

Fig. 4.8 Measure of Efficiency

4.5 Conclusion

In this chapter various aspects of call admission control is considered to obtain a more realistic solution. Here the call admission control problem is divided into three parts. The first we represent the probability of capturing a specific MS in a given network. This probability depends on the velocity and direction of the mobile station as well as its distance from the BS.

Next we represent the feasibility of acceptance of a call by a channel. Here we take care of the interference that may occur in the network. Finally we represent the traffic load that is responsible for optimising load in a system. All three modules combined together the system for the call admission control problem introduced above. Most of the works in this field do not consider all the aspects mentioned above together in a single system but have worked on any one of the aspects at a time.

The methodology of CAC presented here performs better in networks with dynamic parameter settings, such as hotness and feasibility and the load as well. Further, the proposed scheme considers direction and speed of the MS, and the distance of the MS from the BS. The performance of the proposed scheme is compared with existing results on CAC with varying load, and the results are appealing. It is evident from the experimental results that the proposed technique is capable of serving more calls than the other methods, and it can reuse more number of channels than the same by other methods.

The system realised above is a centralized system, where all the calculation and decision making is done by the centralized call admission system and is then broadcasted to all the cells in the network. It is better to use a distributed system, which may perform faster. In such system some of the information should be shared by broadcasting them. The main hurdles in using a fully distributed system are the communication bottleneck. This can, however, be relieved by considering a partly central and partly distributed realization. The trade-off in this case is important to determine which part should remain centralized, and which one to be realized in a distributed manner, so that communication among the system modules is minimized.

References

1. Wang, L., Arunkumaar, S., Gu, W.: Genetic algorithms for optimal channel assignment inmobile communications. In: Proceedings of the 9th International Conference on Neural Information Processing ICONIP 2002, vol. 3, pp. 1221–1225 (November 2002)
2. Hou, J., Fang, Y.: Mobility-based call admission control schemes for wireless mobile networks. Wirel. Commun. Mob. Comput. 1, 269–282 (2001)
3. Sakar, S., Sivarajan, K.N.: Channel assignment algorithms satisfyingm cochannel and adjacent channel reuse constraints in cellular mobile networks. IEEE Trans. Veh. Technol. 51(5), 954–967 (2002)

4. Wang, S.-L., Hou, Y.-B., Huang, J.-H., Huang, Z.-Q.: Adaptive Call Admission Control Based on Enhanced Genetic Algorithm in Wireless/Mobile Network. In: Proceedings of the 18th IEEE International Conference on Tools with Artificial Intelligence (2006)
5. Chen, H., Cheng, C.-C., Yeh, H.-H.: Guard-Channel-Based Incremental and Dynamic Optimization on Call Admission Control for Next-Generation QoS-Aware Heterogeneous Systems. IEEE Transaction on Vehicular Technology 57(5) (September 2008)
6. Thilakawardana, S., Tafazolli, R.: Efficient Call Admission Control and SchedulingTechnique for GPRS Using Genetic Algorithms. Mobile Communications Research Group, Centre for Communications Systems Research, CCSR (2004)
7. Kang, S.H., Sung, D.K.: A CAC Scheme Based on Real-Time Cell Loss Estimation for ATM Multiplexers. IEEE Transaction on Communication 48(2) (February 2000)
8. Xiao, Y., Chen, C.L.P., Wang, Y.: A Near Optimal Call Admission Control With Genetic Algorithm For Multimedia Services In Wireless/Mobile Networks. In: IEEE Nat. Aero. Elec. Conf., pp. 787–792 (October 2000)
9. Deb, K.: Multi-Objective Optimization using Evolutionary Algorithms. John Wiley, London (2001)
10. Nuaymi, L., Godlewski, P., Mihailescu, C.: Call Admission Control Algorithm for Cellular CDMA Systems based on Best Achievable Performance. In: Proc. IEEE Vehic. Tech. Conf. (VT 2000- Spring), Tokyo, vol. 1, pp. 375–379 (2000)
11. Jeon, W., Jeong, D.: Call Admission Control for Mobile Multimedia Communications with Traffic Asymmetry between Uplink and Downlink. IEEE Trans. Vehic. Tech. 50(1), 59–66 (2001)
12. Koo, I., Bahng, S., Kim, K.: Resource reservation in call admission control schemes for CDMA systems with nonuniform traffic distribution among cells. In: Proc. IEEE Veh. Technol. Conf., vol. 1, pp. 438–441 (2003)
13. Tugcu, T., Ersoy, C.: Application of a realistic mobility model to call admissions in DS-CDMA cellular systems. In: Proc. IEEE Veh. Technol. Conf., vol. 2, pp. 1047–1051 (2001)
14. Badia, L., Zorzi, M., Gazzini, A.: On the impact of user mobility on call admission control in WCDMA systems. In: Proc. IEEE Veh. Technol. Conf., vol. 1, pp. 121–126 (September 2002)
15. Ahmed, M.H.: Memorial Call Admission Control In Wireless Networks: A Comprehensive Survey. IEEE Communication Surveys 7(1) (2005)
16. Tian, X., Ji, C.: Bounding the Performance of Dynamic Channel Allocation with QoS Provisioning for Distributed Admission Control in Wireless Networks. IEEE Trans. Vehic Tech. 50(2), 388–397 (2001)
17. Haung, Y., Ho, J.: Distributed Call Admission Control for a Heterogeneous PCS Network Computer. IEEE Trans. Comp. 5(12), 1400–1409 (2002)
18. Islam, M.M., Murshed, M., Dooley, L.S.: A directionally based bandwidth reservation scheme for call admission control. In: 5th International Conference on Computer and Information Technology (ICCIT 2002), Dhaka, Bangladesh, December 26-28 (2002)
19. Wong, D., Lim, T.J.: Soft handoffs in CDMA mobile systems. IEEE Personal Communications Magazine 4(6), 6–17 (1997)
20. Ye, J., Shen, X., Mark, J.W.: Call Admission Control in Wideband CDMA Cellular Networks by Using Fuzzy Logic. IEEE Transactions on Mobile Computing 4(2) (March/April 2005)

21. Chen, Y.-H., Chang, C.-J., Shen, S.: Outage-based fuzzy call admission controller with multi-user detection for WCDMA systems. IEE Proc. Commun. 152(5) (October 2005)
22. Chung-Ju, Chang, L.-C., Kuo, Y.-S., Chen, S.S.: Neural Fuzzy Call Admission and Rate Controller for WCDMA Cellular Systems Providing Multirate Services. In: IWCMC 2006, Vancouver, British Columbia, Canada, July 3–6 (2006)
23. Malarkkan, S., Ravichandran, V.C.: Performance analysis of call admission control in WCDMA System with Adaptive Multi Class Traffic based on Fuzzy Logic. IJCSNS International Journal of Computer Science and Network Security 6(11) (November 2006)
24. Yoshida, T., Watanabe, M., Nishida, S.: Channel prediction for OFDMA using mixtures of experts. International Journal of Knowledge-Based and Intelligent Engineering Systems 10(3), 193–200 (2006)
25. Moura, A.I., Ribeiro, C.H.C., Costa, A.H.R.: WBLS: A signal presence-based Wi-Fi localisation system for mobile devices in smart environments. International Journal of Knowledge-Based and Intelligent Engineering Systems 13(1), 5–18 (2009)
26. Liu, Z., Zarki, M.E.: SIR-Based Call Admission Control for DSCDMA Cellular System. IEEE J. Selected Areas of Comm. 12, 638–644 (1994)
27. Evans, J.S., Everitt, D.: Effective Bandwidth-Based Admission Control for Multiservice CDMA Cellular Networks. IEEE Trans. Vehicular Technology 48, 36–46 (1999)
28. Chang, C., Shen, S., Lin, J., Ren, F.: Intelligent Call Admission Control for Differentiated QoS Provisioning in Wide Band CDMA Cellular System. In: Proc. IEEE Vehicular Technology Conf. (VTC 2000), pp. 1057–1063 (2000)
29. Comaniciu, C., Mandayam, N., Famolari, D., Agrawal, P.: QoS Guarantees for Third Generation (3G) CDMA Systems Via Admission and Flow Control. In: Proc. IEEE Vehicular Technology Conf. (VTC 2000), pp. 249–256 (2000)
30. Bambos, N., Chen, S.C., Pottie, G.J.: Channel Access Algorithms with Active Link Protection for Wireless Communication Networks with Power Control. IEEE/ACM Trans. Networking 8, 583–597 (2000)
31. Andersin, M., Rosberg, Z., Zander, J.: Soft and Safe Admission Control in Cellular Networks. IEEE/ACM Trans. Networking 5, 255–265 (1997)
32. Ho, C.-J., Copeland, J.A., Lea, C.-T., Stüber, G.L.: On Call Admission Control in DS/CDMA Cellular Networks. IEEE Trans. Vehicular Technology 50, 1328–1343 (2001)
33. Fang, Y., Zhang, Y.: Call Admission Control Schemes and Performance Analysis in Wireless Mobile Networks. IEEE Trans. Vehicular Technology 51, 371–382 (2002)
34. Wu, S., Wong, K.Y.M., Li, B.: A Dynamic Call Admission Policy with Precision QoS Guarantee Using Stochastic Control for Mobile Wireless Networks. IEEE/ACM Trans. Networking 10, 257–271 (2002)
35. Fang, Y., Chlamtac, I.: Teletraffic Analysis and Mobility Modelling of PCS Networks. IEEE Trans. Comm. 47, 1062–1072 (1999)
36. Naghshineh, M., Schwartz, M.: Distributed Call Admission Control in Mobile/Wireless Networks. IEEE J. Select. Areas Comm. 14, 711–717 (1996)
37. Dziong, Z., Jia, M., Mermelstein, P.: Adaptive Traffic Admission for Integrated Services in CDMA Wireless-Access Networks. IEEE J. Seleced Areas of Comm. 14, 1737–1747 (1996)
38. Klir, G.J., Yuan, B.: Fuzzy Sets and Fuzzy Logic: Theory and Applications. Prentice-Hall (1995)
39. Wang, L.X.: A Course in Fuzzy Systems and Control. Prentice Hall (1997)

40. Rappaport, T.S.: Wireless Communications: Principles and Practice. Prentice Hall (1996)
41. Viterbi, A.J., Viterbi, A.M., Gilhousen, K.S., Zehavi, E.: Soft Handoff Extends CDMA Cell Coverage and Increase Reverse Link Capacity. IEEE J. Selected Areas of Comm. 12, 1281–1288 (1994)
42. Braae, M., Rutherford, D.A.: Fuzzy Relations in a Control Setting. Kybernetes 7, 185–188 (1978)
43. Del Re, E., Frantacci, R., Giambene, G.: Handover and Dynamic Channel Allocation Techniques in Mobile Cellular Networks. IEEE Trans. Vehicular Technology 44, 229–237 (1995)
44. Liu, T., Bahl, P., Chlamtac, I.: Mobility Modelling, Location Tracking, and Trajectory Prediction in Wireless ATM Networks. IEEE J. Selected Areas Comm. 16, 922–936 (1998)
45. Guerin, R.A.: Channel Occupancy Time Distribution in a Cellular Radio System. IEEE Trans. Vehicular Technology 36, 89–99 (1987)
46. Seskar, I., Maric, S., Holtzman, J., Wasserman, J.: Rate of Location Area Updates in Cellular Systems. In: Proc. IEEE Vehicular Technology Conc. (VTC 1992), pp. 694–697 (1992)
47. Nanda, S.: Teletraffic Models for Urban and Suburban Microcells: Cell Sizes and Handoff Rates. IEEE Trans. Vehicular Technology 42, 673–682 (1993)
48. Varshney, U., Jain, R.: Issues in emerging 4G wireless networks. IEEE Computer 34(6), 94–96 (2001)

Chapter 5
Call Admission Control Using Bio-geography Based Optimization

The chapter proposes a new approach to call admission control in a mobile cellular network using Bio-geography based optimization. Existing algorithms on call admission control either ignore both variation in traffic conditions or velocity of mobile devices, or at most consider one of them. This chapter overcomes the above problems jointly by formulating call admission control as a constrained optimization problem, where the primary objective is to minimize the call drop under dynamic condition of the mobile stations, satisfying the constraints to maximize the channel assignment and minimize the dynamic traffic load in the network. The constrained objective function has been minimized using Bio-geography based optimization. Experimental results and computer simulations envisage that the proposed algorithm outperforms most of the existing approaches on call admission control, considering either of the two issues addressed above.

5.1 Introduction

Call Admission Control (CAC) refers to the problem of efficient call management in a mobile cellular network. The primary objective of CAC is to serve as many calls as possible, and prevent dropping of calls in progress [1-3]. Besides this, an efficient call management also aims at satisfying additional (secondary) objective to assign appropriate channels to the incoming/handoff calls, so that the necessary soft constraints for channel assignment are maintained [1-8]. Typically, soft constraints include co-channel, co-site, and adjacent channel constraints, all of which need to be satisfied to serve the secondary objective. In the current literature, Quality of Service (QoS) is often used to measure the quality of CAC with an attempt to maximize call assignment and soft handoff, satisfying the soft constraints. The better the QoS, the better is the CAC.

The concept of biogeography can be traced to the work of nineteenth century naturalists such as Alfred Wallace and Charles Darwin. Robert Macarthur and Edward Wilson began working together on mathematical models of biogeography in 1960. The primary objective was on the distribution of species among neighboring islands. The mathematical models were developed for the extinction and migration of species. The application of biogeography to engineering is similar to what has occurred in the past few decades with genetic algorithms (GA),

S. Ghosh and A. Konar: CAC in a Mobile Cellular Network, SCI 437, pp. 153–181.
springerlink.com   © Springer-Verlag Berlin Heidelberg 2013

neural networks, fuzzy logic, particle swarm optimization (PSO), and other areas of computational intelligence.

Biogeography-based optimization (BBO) rests on the migration strategy of animals to solve the problem of global optimization. In general, Biogeography is the study of the geographical distribution of biological organisms. Mathematical equations that govern the distribution of organisms were first discovered and developed during the 1960s. The researchers can learn from nature and it motivates the application of biogeography to optimization problems. This chapter considers the mathematics of biogeography as the basis for the development of a new field for admission of calls in a mobile network using the biogeography-based optimization (BBO).

The BBO migration strategy is similar to the global recombination approach of evolutionary strategies (ES) (Black, 1996), (Black et al., 1997), in which many parents can contribute to a single offspring. Global recombination has also been adapted to GA (Eiben, 2003), (Eiben, 2000), but BBO differs from GAs in one important aspect. In GA recombination is used to create new solutions, while in BBO migration is used to change existing solutions.

Global recombination in ES is a reproductive process, which creates new solutions, while BBO migration is an adaptive process that modifies existing solutions. A quantitative comparison between BBO and other Evolutionary Algorithms is included in Simon (2008), where 14 benchmark functions, each with 20 dimensions, were studied. It was shown that BBO and the stud GA (so named for its selection of the best individual in the population as one of the parents for every crossover operation) performed the best out of eight EA.

5.1.1 The Problem

CAC systems are mainly used to take decisions, whether a call should be serviced, blocked or dropped by a Base Station (BS), and if serviced, it identifies the channel to be assigned to that call. At the time of taking such decision, the interference is considered the only factor in most of the current literature [1 - 48].

Existing works presented above, concentrates on specific aspects of the call admission control problem. A few of these considers rectangular cells and the near neighbor and hence ignores the entirety of the network, while the rest considers only the hand off and call drop due to insufficient channels. Unfortunately, most of the above works consider the mobile stations either static or moving very slowly in the small area with very low traffic load.

The mobility factor comprising speed and direction of movement has been ignored in these works. Moreover, the overall network load is also ignored and hence the handoff policies used have a partial effect on the network. Most of the above schemes are localized to a single cell. Hence the channel reuse is also not done efficiently.

In this chapter, we propose a scheme that takes a more wide view of the call admission control problem. We consider our cells to be hexagonal so as to easily track the movement of MS in the neighborhood cells. Instead of considering a single cell scenario, we have taken a small network to implement the algorithm to incorporate the intercellular communication efficiently. The decision of acceptance or rejection of a call dose not only depend on the feasibility and availability of the channel but also on the speed at which the MS moves and its direction of movement. Its geographical location with respect to the base station also has a significant importance. Moreover the traffic density on a cell is also considered to be a determining factor. The reuses of the channels also make the approach more effective. The algorithm proposed in the chapter is based on all the factors mentioned above which more perfectly handles the real world situation.

There are certain parameters to measure the efficiency of such a system. One important parameter is the number of call drop that shows the number of unwanted call disconnection in the system and should be as low as possible. The other one would be the call block, where the new call attempts are rejected by the network due to the insufficiency of the free channels. The number of assignable channels that helps in increasing the efficiency of the system is another important parameter considered here.

In this chapter, we propose a scheme, which takes into account the motion aspect of the mobile station (MS) as well, besides considering all the necessary objectives of CAC. We here use three parameters: i) speed, ii) direction, and iii) distance of the MS from the nearest base station to model the motion of the mobile station. The above three parameters play an important role at the time of soft handoff of a call from one cell to other. The importance can be explained with the help of Fig.5.1, where the central cell has six neighbours.

The MS in such a cell while in service may move in various directions with different speed. If it moves toward cell 5 or 6 directly then the channel for handoff will be searched in those cells. When it moves slowly along the common boundary of cell 5 and 6, the cell with base station nearer to the current location of MS is considered. Again, if the above movement takes place with a very high speed, then the call may be dropped due to high interference. Suppose, it moves towards cell 2 very slowly, then there may not be a requirement for a soft handoff at all since it may never cross the existing cell boundary.

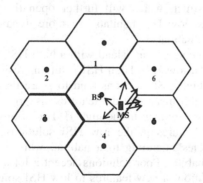

Fig. 5.1 Illustrating the movement of MS in a mobile cellular network

Hence, we aim at searching an optimal set of assignment, which will take care of the quality of service, and the velocity aspect of the scheme.

5.2 The BBO Approach

Mathematical models of biogeography explain the migration of species from one island to another and their evolution and extinction [49-52]. The term "island" is used descriptively rather than literally. An island is any habitat that is geographically isolated from other habitats. The more generic term "habitat" will be used (rather than "island"). Geographical areas that are well suited as residences for biological species are said to have a high suitability index variables (SIV). Features that correlate with HSI include such factors as rainfall, diversity of vegetation, diversity of topographic features, land area, and temperature. The variables that characterize habitability are called suitability index variables (SIV). SIV can be considered the independent variables of the habitat, and HSI can be considered the dependent variable.

Habitats with a high HSI tend to have a large number of species, while those with a low HSI have a small number of species. Habitats with a high HSI have many species that emigrate to nearby habitats, simply by virtue of the large number of species that they host. Habitats with a high HSI have a low species immigration rate because they are already nearly saturated with species. Therefore, high HSI habitats are more static in their species distribution than low HSI habitats. The high HSI habitats have a high emigration rate; the large number of species on high HSI islands has many opportunities to emigrate to neighboring habitats.

This does not mean that an emigrating species completely disappears from its home habitat; only a few representatives emigrate, so an emigrating species remains extant in its home habitat, while at the same time migrating to a neighboring habitat.) Habitats with a low HSI have a high species immigration rate because of their sparse populations.

This immigration of new species to low HSI habitats may raise the HSI of the habitat, because the suitability of a habitat is proportional to its biological diversity. However, if a habitat's HSI remains low, then the species that reside there will tend to go extinct, which will further open the way for additional immigration. Due to this, low HSI habitats are more dynamic in their species distribution than high HSI habitats.

A good solution is analogous to an island with a high HSI, and a poor solution represents an island with a low HSI. High HSI solutions resist change more than low HSI solutions. The high HSI solutions tend to share their features with low HSI solutions. This does not mean that the features disappear from the high HSI solution; the shared features remain in the high HSI solutions, while at the same time appearing as new features in the low HSI solutions. This is similar to representatives of a species migrating to a habitat, while other representatives remain in their original habitat. Poor solutions accept a lot of new features from good solutions. This addition of new features to low HSI solutions may raise the quality of those solutions.

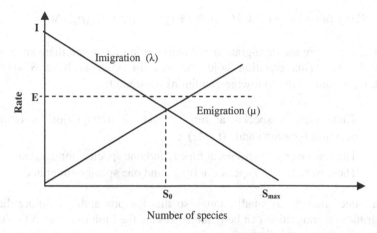

Fig. 5.2 Species model of a single habitat based on biogeography

The model of species abundance in a single habitat is shown in Fig. 5.2. The immigration rate λ and the emigration rate μ are functions of the number of species in the habitat.

For the immigration curve, the maximum possible immigration rate to the habitat is I, which occurs when there are zero species in the habitat. As the number of species increases, the habitat becomes more crowded, fewer species are able to successfully survive immigration to the habitat, and the immigration rate decreases. The largest possible number of species that the habitat can support is S_{max}, at which point the immigration rate becomes zero.

For the emigration curve if there are no species in the habitat then the emigration rate must be zero. As the number of species increases, the habitat becomes more crowded; more species are able to leave the habitat to explore other possible residences, and the emigration rate increases. The maximum emigration rate is E, which occurs when the habitat contains the largest number of species that it can support.

The equilibrium number of species is S_o, at which point the immigration and emigration rates are equal. However, there may be occasional excursions from due to temporal effects. Positive excursions could be due to a sudden spurt of immigration (caused, perhaps, by an unusually large piece of flotsam arriving from a neighboring habitat), or a sudden burst of speciation (like a miniature Cambrian explosion). Negative excursions from could be due to disease, the introduction of an especially ravenous predator, or some other natural catastrophe. It can take a long time in nature for species counts to reach equilibrium after a major perturbation.

The immigration and emigration curves in shown in Fig. 5.2 as straight lines but, in general, they might be more complicated curves.

Now, the probability Ps is the habitat contains exactly S species. Ps changes from time t to time ($t + \Delta t$) as follows:

$$P_s(t+\Delta t) = P_s(t)(1 - \lambda_s \Delta t - \mu_s \Delta t) + P_{s-1}\lambda_{s-1}\Delta t + P_{s+1}\mu_{s+1}\Delta t \qquad (5.1)$$

where λ_s and μ_s are the immigration and emigration rates when there are S species in the habitat. This equation holds because in order to have S species at $(t + \Delta t)$ time, one of the following conditions must hold:

- There were S species at time t, and no immigration or emigration occurred between t and $(t + \Delta t)$;
- There were (S - 1) species at time t, and one species immigrated;
- There were (S + 1) species at time , and one species emigrated.

It is assumed that Δ t is small enough so that the probability of more than one immigration or emigration can be ignored. Taking the limit of 5.1 as $\Delta t \rightarrow 0$ gives equation 5.2 shown as follows:

$$\overset{\wedge}{P_S} = \begin{cases} -(\lambda_s + \mu_s)P_s + \mu_{s+1}P_{s+1} & S = 0 \\ -(\lambda_s + \mu_s)P_s + \lambda_{s-1}P_{s-1} + \mu_{s+1}P_{s+1}, & 1 \leq S \leq S_{max} - 1 \\ -(\lambda_s + \mu_s)P_s + \lambda_{s-1}P_{s-1} & S = S_{max} \end{cases} \qquad (5.2)$$

Say, $n = S_{max}$ and $P = [P_0 P_1 P_2 \ldots P_n]^T$

Now, we can arrange the equations of equation 5.2 into the single matrix equation

$$\overset{\circ}{P} = AP \qquad (5.3)$$

Where the matrix A is given in the following equation:

$$A = \begin{bmatrix} -(\lambda_0 + \mu_0) & \mu_1 & 0 & \ldots & & 0 \\ \lambda_0 & -(\lambda_1 + \mu_1) & \mu_2 & \ldots & & \ldots \\ \ldots & \ldots & \ldots & \ldots & & \ldots \\ \ldots & \ldots & \lambda_{n-2} & -(\lambda_{n-1} + \mu_{n-1}) & \mu_n \\ 0 & \ldots & 0 & \lambda_{n-1} & -(\lambda_n + \mu_n) \end{bmatrix} \qquad (5.4)$$

For the straight-line curves shown in Fig.5.3 we have

$$\mu_k = \frac{Ek}{n}$$

$$\lambda_k = I\left(1 - \frac{k}{n}\right) \qquad (5.5)$$

Now for special case $E = I$, then

$$\lambda_k + \mu_k = E \tag{5.6}$$

According to the simplified form stated in equation 5.6, the species model will be the following type.

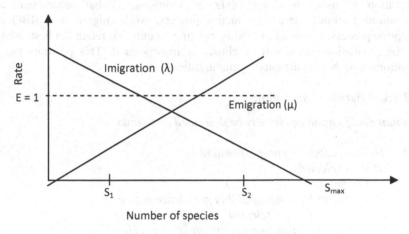

Fig. 5.3 S_1 is relatively a poor solution and S_2 relatively a good solution

5.2.1 Migration in BBO

Suppose that we have a problem and a population of candidate solutions that can be represented as vectors of integers. Each integer in the solution vector is considered to be an SIV. The assessment for the goodness of the solutions has to be done. The solutions that are good are considered to be habitats with a high HSI, and those that are poor are considered to be habitats with a low HSI. HSI is analogous to "fitness" in other population-based optimization algorithms (GAs, for example). High HSI solutions represent habitats with many species, and low HSI solutions represent habitats with few species. The identical species curve ($E = I$) is considered for simplicity but the S value represented by the solution depends on its HSI.

S_1 in Fig.5.3 represents a low HSI solution, while S_2 represents a high HSI solution. S_1 in Fig.5.3 represents a habitat with only a few species, while S_2 represents a habitat with many species. The immigration rate λ_1 for S_1 will be higher than the immigration rate λ_2 for S_2. The emigration rate $\mu 1$ for S_1 will be lower than the emigration rate $\mu 2$ for S_2. The emigration and immigration rates of each solution probabilistically share information between habitats. With probability P_{mod}, each solution is modified based on other solutions.

If a given solution is selected to be modified, then the immigration rate λ to probabilistically decide whether or not to modify each suitability index variable (SIV) in that solution. If a given SIV in a given solution S_i selected to be modified, then the emigration rates μ of the other solutions to probabilistically decide which of the solutions should migrate a randomly selected SIV to solution S_i.

The BBO migration strategy is similar to the global recombination approach of the breeder GA and evolutionary strategies in which many parents can contribute to a single offspring, but it differs in at least one important aspect. In evolutionary strategies, global recombination is used to create new solutions, while BBO migration is used to change existing solutions. Global recombination in evolutionary strategy is a reproductive process, while migration in BBO is an adaptive process; it is used to modify existing islands. To retain the best solutions in the population, some sort of elitism is incorporated. This prevents the best solutions from being corrupted by immigration.

5.2.1.1 Migration Algorithm

Habitat modification can loosely be described as follows:

Select H_i with probability proportional to λ_i
    If H_i is selected
       For j=1 to n
          Select H_j with probability proportional to μ_j
            If H_j is selected
               Randomly select an SIV from H_j
               Replace a random SIV in with
      end
    end
end

5.2.2 Mutation in BBO

A habitat's HSI can change suddenly due to apparently random events (unusually large flotsam arriving from a neighboring habitat, disease, natural catastrophes, etc.) The model of BBO as SIV mutation, and species count probabilities is used to determine mutation rates.

The probabilities of each species count will be governed by the differential equation given in 5.2. By looking at the equilibrium point on the species curve of Fig.5.2, it is observed that low species counts and high species counts both have relatively low probabilities and medium species counts have high probabilities because they are near the equilibrium point.

Each population member has an associated probability, which indicates the likelihood that it was expected *a priori* to exist as a solution to the given problem. Very high HSI solutions and very low HSI solutions are equally improbable. Medium HIS solutions are relatively probable. If a given solution S has a low probability P_s, then it is surprising that it exists as a solution. It is, therefore, likely to mutate to some other solution. Conversely, a solution with a high probability is less likely to mutate to a different solution. The mutation rate that is inversely proportional to the solution probability,

$$m_i = m_{\max}\left(1 - \frac{P_i}{P_{\max}}\right)$$

Where

m_{max} is a user-defined parameter,

and $\quad P_{max} = argmax\, P_i$, $i = 1,...NP$.

This mutation scheme tends to increase diversity among the population. Without this modification, the highly probable solutions will tend to be more dominant in the population. This mutation approach makes low HSI solutions likely to mutate, which gives them a chance of improving. It also makes high HSI solutions likely to mutate, which gives them a chance of improving even more than they already have. Note that we use an elitism approach to save the features of the habitat that has the best solution in the BBO process, so even if mutation ruins its HSI, we have saved it and can revert back to it if needed. So, we use mutation (a high risk process) on both poor solutions and good solutions. Those solutions that are average are hopefully improving already, and so we avoid mutating them (although there is still some mutation probability, except for the most probable solution).

Mutation Algorithm: Mutation can be described as follows:
For j=1 to m
    Use λ_i and μ_i to compute the probability P_i
        Select SIV H_i (j) with probability proportional to P_i
           If H_i (j) is selected
               Replace H_i (j) with a randomly generated SIV
           end
end

5.3 Formulation

Here we have attempted to handle the calls with their mobility factor too. A caller may be in a move when a call is generated and while serviced. But depending on the speed of the MS and the direction in which they are moving, there may be a need to reassign the call to a different channel of a different cell. To explain such situation we start with defining the primitives.

5.3.1 Definitions

We consider a system of M hexagonal cells present in the network and each of them has N number of frequency channels. In CAC we need to find out the best allocation of calls in different cells. So we need to know about the current status of allocation, which is usually represented by an allocation matrix. In this chapter, we plan to select the appropriate allocation matrix online, so as to satisfy given objective function and systems constraints to be introduced later. The formal definition of allocation matrix is given here for convenience.

Definition 1: Let $[f_{m,\,i}], \forall i \in M, m \in N$, be a binary matrix describing allocation of channels in given cells, where

$$F = [f_{m,i}] = \begin{cases} 1, & \text{if } m^{th} \text{ channel is allocated to serve a call in the } i^{th} \text{ cell} \\ 0, & \text{if the } m^{th} \text{ channel in the } i^{th} \text{ cell is free.} \end{cases}$$

Example5.1: An example of a 4x3 allocation matrix where there are 4 cells and 3 channels is given as

$$channel \rightarrow$$

$$Cell \downarrow \quad \begin{bmatrix} 0 & 1 & 0 \\ 0 & 0 & 0 \\ 1 & 0 & 0 \\ 0 & 0 & 1 \end{bmatrix}$$

This implies that the 2nd channel of the 1st cell is serving a call and rest of the channels are free. The knowledge of calls assigned to channels in any cell is a very important to measure the feasibility of assigning the new calls satisfying the soft constraints.

Definition 2: Let *assignment matrix* $A = [a_{m,i}], \forall i \in M, m \in N$, be a matrix describing the assignment of calls to the channels of a cell, where

$$a_{m,i} = p, \quad \text{where } p \text{ is the number of channel assigned to the } m^{th} \text{call in cell } i$$

Example5.2: Let there be 10 calls in the network with 4 cells and 3 channels. Then

$$calls \rightarrow$$

$$Cell \downarrow \quad \begin{bmatrix} 1 & 0 & 0 & 0 & 2 & 0 & 0 & 0 & 0 & 3 \\ 0 & 0 & 1 & 0 & 0 & 0 & 0 & 0 & 2 & 0 \\ 0 & 2 & 0 & 0 & 0 & 1 & 0 & 3 & 0 & 0 \\ 0 & 0 & 0 & 1 & 0 & 0 & 3 & 0 & 0 & 0 \end{bmatrix}$$

We can see that the 1st call is served by the 1st channel of the 1st cell, the 2nd call by the 2nd channel of 3rd cell, 3rd call by the 1st channel of 2nd cell and so on.

While allocating the calls to the channels we should maintain the QoS in terms soft constraints. The soft constrains ensure that the new call assignment do not have any interference with the existing calls assignment in the neighbouring cells or in the same cell. In this chapter, we measure the QoS as a function of three important network attributes: feasibility, hotness and motion of the MS. The feasibility of channel assignment is often expressed as linear combination of allocation and compatibility matrices. A formal definition of compatibility matrix is given here.

Definition 3: The *compatibility matrix C* gives a measure of satisfaction of the soft constraints, attempted to minimize co-channel, co-site and the adjacency-channel interference, whose non-diagonal and diagonal elements are expressed by

$$C_{j,k} = \min imum \ chann \ el \ seperat \ ion \ requir \ ed \ for \ a \ c \ all \ assign \ ment$$

$$in \ cell \ j \ when \ th \ ere \ are \ ca \ lls \ assign \ ed \ in \ cell \ k, and$$

$$C_{i,i} = minimum \ channel \ seperation \ required \ to \ assign \ a \ call \ in \ cell \ i$$

$$when \ there \ are \ other \ channels \ assigned \ in \ the \ same \ cell.$$

Speed of the MS is a very important aspect of the CAC since it affects the soft handoff and the call drop process.

Example 5.3: A compatibility matrix in a 4 cell network is given as

$$\begin{bmatrix} 4 & 3 & 2 & 0 \\ 3 & 4 & 3 & 2 \\ 2 & 3 & 4 & 3 \\ 0 & 2 & 3 & 4 \end{bmatrix} \begin{array}{l} where, \ C_{i,i} = 4 \ and \\ \\ C_{i,j} = \{0 \ to \ 3\} \end{array}$$

Definition 4: *Speed* $V=[v_{p,i}]$ in the present context refer to rate of position changing of a MS busy with a call utilizing a channel p in cell i.

Example 5.4:

$$v_{p,i} = 0 \qquad when \ the \ MS \ is \ static$$

$$0 < v_{p,i} \leq 60 \qquad when \ inside \ the \ city$$

$$v_{p,i} > 60 \qquad in \ highways$$

Distance of a MS from the BS is also an important aspect as it takes part in the soft handoff process.

Definition 5: Distance of MS j from the BS i $\forall i \in M$ is denoted as
$$\textbf{\textit{Dis}} = [dist_{i,j}].$$

Example 5.5: $dist_{i,j} = d$ where d is the distance of the MS from BS as in Fig 5.4

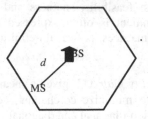

Fig. 5.4 Distance between BS and MS

Hotness is another major factor while deciding about a call assignment to a channel. If the number of incoming calls becomes very high the limited resource of the system will not be able to handle the incoming calls and thus subsequent new calls will be blocked.

Definition 6: Hotness of a cell is defined as the number of incoming calls per unit time and is denoted as H= $[h_i]$.

Example5.6: Hotness in a 4-cell scenario where maximum number of incoming calls is 10 can be given as

$$H = [h_i] = \begin{bmatrix} 2 & 10 & 8 & 5 \end{bmatrix}$$

Angle of motion of MS with respect to the BS shows the direction in which the MS is moving and hence the search for cells with free channels becomes easier. Hence it is also an important factor in CAC.

Definition 7: Angle of motion is the angle made by the direction of motion of a MS p with respect to the BS i and is denoted by ***Delt*** = $[\delta_{p,i}]$, and is illustrated in Fig.5.5.

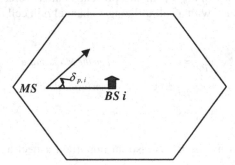

Fig. 5.5 Angle of Motion

Time taken by each call is useful for finding out the calls going for a very long time. At the time of high congestion, these calls are dropped to free the channels for reuse.

5.3.2 Formulation

Let the time taken by each call in a given channel p, of each cell i is denoted by $T_{p,i}$. If $v_{p,i}$ be the speed with which an MS is moving in any direction, then, its velocity along that direction is $v_{p,i} \cos \delta_{p,i}$. Again if $dist_{p,j}$ be the distance traversed by the MS in time $\Delta T_{p,i}$ then average probability of capturing that MS is denoted as P_{CMS} and is given by

$$P_{CMS} = \sum_p \frac{(v_{p,i} \cos \delta_{p,i}) \Delta T_{p,i}}{dist_{p,j}} \tag{5.7}$$

An increase in the value of the above expression is created by a high range of velocity, which gives very little time to search a new channel and to go for a soft hand off (SHO). Hence it should be minimized.

The difference between the two calls in the two different channels should be at a minimum distance to avoid the interference described as soft constrains above. The feasibility of assignment of calls in a cell i can be checked by satisfying inequality (1)[1],[3].The condition demands that the channel separation should satisfy the bare minimum value obtained from compatibility matrix and is denoted by Feas.

$$\sum_m \sum_n (\sum_i (|a_{m,i} - a_{n,j}|)) < \sum_i C_{i,j} \tag{5.8a}$$

$$\tag{5.8b}$$

$$\Rightarrow Feas^j = \sum_m \sum_n (\sum_i (|a_{m,i} - a_{n,j}| - C_{i,j})) < 0$$

It follows from the last inequality that smaller the value of the left hand side, lesser is the interference and thus better is the quality of service of the call assigned to the channel.

The traffic load in a cell is an important issue to determine the possible admission of incoming calls in a given cell. Traffic load in the cell i may be expressed as the ratio of incoming calls and the total free channels of the cell. We define a metric to measure the traffic load in a given cell i, denoted as Load and is given by

$$Load = \frac{h_i}{\sum_m (1 - f_{m,i})} \tag{5.9}$$

Hence the traffic load is the more if the number of incoming calls exceeds considerably than the free cells in the system and starts affecting the overall system performance. Hence this is also to be minimized.

We now construct an objective function, the minimization of which yields a possible solution to the call admission problem. Since expressions (5.7), (5.8b) and (5.9) all need to be minimized, a minimization of their linear combination offers an objective that jointly satisfies all the three basic objectives. The overall objective function is given by

$$Z_{habitat} = P_{CMS} + Feas + Load$$

$$= \sum_i \left(\sum_p \frac{(v_{p,i} \cos\delta_{p,i}) \Delta T_{p,i}}{dist_{p,i}} + \sum_m \sum_n (\sum_j (|a_{m,i} - a_{n,j}| - C_{i,j})) + \frac{h_i}{\sum_m (1 - f_{m,i})} \right) \qquad (5.10)$$

The lower the value of the function better is the performance of CAC system.

Here we also define the difference of fitness ΔZ as

$$\Delta Z = Z_{offspring} - Z_{parent} \qquad (5.11)$$

5.4 CAC Realized with BBO

In the above formulated problem we consider the networks as **habitats** with suitability index variables (SIV) as the velocity, distance , angle, the hotness of each cell and the feasibility of assignment of the channels. The suitability index variables (SIV) of each habitat is given by $Z_{habitat}$. Here the calls waiting for service are said to be the **species** of the habitat. the maximum possible immigration rate to the habitat I is the maximum number of free channels and the maximum emigration rate E in this problem is the number of channels in service respectively.

Let M, N *Total number of channel and cell;*
      ITE   *Total number of iteration;*
      L   *Long call interval;*
      ε   *Very small number;*
      s   *Size of population;*
      V   *Speed as given in definition 4*
      H   *Hotness as given in definition 6*
      A   *Assignment matrix as given in definition 2*
      C   *Compatibility matrix as given in definition 3.*
    Delt   *Angle of motion as given in definition 7*
    Dis   *Distance as given in definition 5.*
  $P = \{p^j\}$ *Initial population.*
  C-Pool   *Parent Pool.*
  O-Pool   *Offspring Pool*
      Z *fitness for a candidate*

Here, we propose a CAC algorithm which ensures minimization of call drop and noise. We use BBO to ensure the optimization. We first consider a pool of solutions of represented as assignment matrix. We make sure all of the initial solutions are unique. Then following the BBO steps, we select a pair of solutions by a roulette wheel. We use the pair of solutions as an input to the BBO migration.

The migration process has two parts *a*) rearrange *b*) change. In 'rearrange' we identify a random point common to both the solution and perform exchange in the two. This causes the generation of a new set of solution from the existing one. In change, we consider the random hang-up of calls and change the rearranged solution and the parent solutions accordingly.

After this we go for a fitness check to see whether the new solutions are viable or not. Depending on the measured fitness we accept or reject the new solutions.

At the end to avoid the congestion in the network, we drop some calls forcefully by defining a concept call long call. If the call connection exceeds certain duration, we consider them as long calls. The whole process is repeated as long as the fitness difference between the parent and the offspring is considerably low.

The algorithm is formally given below.

Algorithm:
CALL_Adm(A,C;Z)
   BEGIN
      FOR (j=0 ; j<s; j++)
       Initial (A, C; p^j) ;
        END FOR.
      FOR (j=0 ; j<s; j++)
       R_Dup(p^j; p^j) ;
   END FOR
   P={p^j};
   Initialize S_{max}, E, I.
FOR (i =0; i<ITE; i++)
      Do

        Selection (A, C, P; C-Pool);
       Migration (C-Pool; O-Pool);
       FOR (j=0; j<s; j++)
        Fit (A, C, Dist, Delt , V, H, C-Pool ; Z_{Pool}) ;
        Fit (A, C, Dist, Delt , V, H, O-Pool ; Z_{Pool}) ;
       Accept (C-Pool, O-Pool, O-Pool),
     END FOR.
     ($T_{x,i,j}$)++;

     IF ($T_{x, i,j}$ > L)

   Drop_call (x, i, j ,O-Pool; O-Pool);   *END IF;*
      IF (ΔZ_f^j <ε)
    BREAK;
    END IF;
   WHILE (CONDITION 8a)
   END FOR;
END.

Initialization:

In this chapter, we consider the habitats as the assignment of calls to channels of all the cells in the network. Hence we have each solution as a M x N vector F (assignment) such that

$$F = \{f_{m,i}\} = \begin{cases} 1 \text{ when } m^{th} \text{ channel in } i^{th} \text{ cell is in service} \\ 0 \text{ otherwise} \end{cases},$$

where F satisfy the soft constraints. We have considered a pool of p such vectors of assignment as a parent population all of which satisfies the soft constraints.

Initial (A, C; p^j)
Step1. Loop i (0, N-1), 1.
Step2. Loop m (0, M-1), 1.
Step3. Check condition stated in (8a)
Step4. If true make $f_{m,i} = 1$,   else 0
Step5. Goto step 1.
Step6. Construct $p^j = F = [f_{m,i}]$
Step7. End.

Example5.7: We consider a network with 3 cells and 4 channels. Then the parent matrix, obeying the compatibility constraints, may look like the following p^j where a '1' represents a channel in a cell is assigned to a call and a '0' represents a free channel.

$$p^j = F = [f_{m,i}] = \begin{bmatrix} 0 & 1 & 0 \\ 0 & 0 & 0 \\ 0 & 0 & 1 \\ 0 & 0 & 0 \end{bmatrix} \text{ where, } M=3 \text{ and } N=4.$$

Remove / Replace Duplicates

This step is done by comparing the habitats and replacing the duplicates so that no species is used against itself.

Fitness Evaluation

This is the part where we calculate the fitness of habitats using the equation (5.4) where all the other values are known.

Fit (A, C, Dist, Delt , V, H, Pool,; Z $_{pool}$) :
For all the habitats of the pool we first find out the individual

$$P^i_{CMS} = \sum_p \frac{(v_{p,i} \cos \delta_{p,i}) \Delta T_{p,i}}{dist_{p,i}}$$

Then we calculate the feasibility of a species using

$$P^i_{CMS} = \sum_p \frac{(v_{p,i} \cos \delta_{p,i}) \Delta T_{p,i}}{dist_{p,i}}$$

At the end we calculate lode for all the habitat using $Load^i = \dfrac{h_i}{\sum\limits_m (1 - f_{m,i})}$

Together $\sum\limits_i P^i_{CMS} + Feas^i + Load^i$ *the total fitness is calculated.*

Selection

A roulette wheel with slots sized according to feasibility is used for selection process. We construct such a roulette wheel as follows

- First we calculate the feasibility value for a parent p^j using condition (5.8b) denoted as *feas*^{j}.

- Then we calculate the total feasibility of all parents $\sum\limits_{j=0}^{s-1} feas^j$.

- Now we calculate the probability of selection of a parent p^j as

$$\Pr ob_{p^j} = \frac{feas^j}{\sum\limits_{j=0}^{s-1} feas^j}.$$

- Now we calculate the cumulative probability $CP_i = \sum\limits_{j=0}^{i} \Pr ob_{p^j}$.

- The selection process is based on spinning the roulette wheel s times; each time we select a single chromosome for a new population in the following way:
    - Generate a random (floating) number r in the range [0; 1],
    - If $r < CP_1$ then select the 1st chromosome (p^1);
    - Otherwise select the ith chromosome p^i $(2 < i < s)$ if $PROB_{i-1} < r < PROB_i$.

This way we can choose the parents who have higher feasibility of getting new calls in the population.

Migration

The migration algorithm is given as

Migration (A, C, P, C-Pool, O-Pool)
FOR (i =0;i<ITE; i++)
    Do

              Rearrange (C-Pool; O-Pool);
                    Change (O-Pool; O-Pool);
        END FOR

Where the functions are given as follows:

Rearrange

We use the C-Pool generated by the selection procedure for rearrange. We consider two random chromosomes from the C-Pool to rearrange. We consider a one-point rearrange and generate a random position (r, c) at which the rearrange takes place to produce two offspring.

Let $p_{m,l}^{x}$ and $p_{m,l}^{y}$ be two parents taking part in rearrange. After rearrange the offspring are kept in O-Pull.

Example5.8: Let the parents have 3 cells and 4 channels and the rearrange point is randomly chosen as 3rd row 3rd column.

$$\text{i.e } r_{max} = 4 \text{ and } c_{max} = 3$$

$$p^1 = \begin{matrix} 0 & 1 & 0 \\ 0 & 0 & 0 \\ 1 & 0 & 0 \\ 0 & 0 & 1 \end{matrix} \qquad p^2 = \begin{matrix} 1 & 0 & 0 \\ 0 & 0 & 1 \\ 0 & 0 & 0 \\ 0 & 1 & 0 \end{matrix}$$

Then after rearrange at r = 3 and c =3 the offspring become

$$f^1 = \begin{matrix} 0 & 1 & 0 \\ 0 & 0 & 0 \\ 0 & 0 & 0 \\ 0 & 1 & 0 \end{matrix} \text{ and } f^2 = \begin{matrix} 1 & 0 & 0 \\ 0 & 0 & 1 \\ 1 & 0 & 0 \\ 0 & 0 & 1 \end{matrix}$$

Change:

We consider the random call hang-up in the system and express this phenomenon as Change. Suppose in a given cell r, a call served with channel c is disconnected by the caller. Then following the definition of allocation matrix F, we understand that the element $f_{ml} = 0$ for m= r and l=c, after disconnection of the call.

$$\text{i.e. } f_{m,l} = 0, \text{ where } m = r, l = c$$

We have considered minimum two such hang-ups in the network. Accordingly random positions are generated where Change is done as described above.

Example5.9: Let there be Change at the positions (1, 2) and (4, 3) for the parents with 4 cells and 3 channels

$$\text{i.e., } r_{max} = 4 \text{ and } c_{max} = 3 \text{ and}$$

$$f = \begin{matrix} 0 & 1 & 0 \\ 0 & 0 & 0 \\ 1 & 0 & 0 \\ 0 & 0 & 1 \end{matrix}$$

Now, r1= 1, c1=2, and r2 = 4, c2 = 3. Hence after Change the offspring will become

$$f = \begin{matrix} 0 & \boxed{0} & 0 \\ 0 & 0 & 0 \\ 1 & 0 & 0 \\ 0 & 0 & \boxed{0} \end{matrix}$$

Call Drop: this is where the call is forcefully terminated due to the time constraint.

***Drop_call** (x, i, O-Pool; O-Pool)*
    BEGIN

$$f_{i,j}^{x} = 0;$$

    END.

Accept:

Here we combine both the C-Pool, O-Pool together and then decide which habitat are to be retained and who are discarded. Depending on the SVI we sort the habitats and retain the habitats having SVI greater than a given value of SVI remaining we discard.

    ***Accept**(C-Pool, O-Pool, O-Pool)*
        BEGIN
            FOR (j=0; j<2 S_{max} , j++)
                Z^{j} = Sort ($Z_{O\text{-}Pool}$, $Z_{C\text{-}Pool}$),

> *IF(j>Smax)*
> *Keep p^j // immigrate*
> *END FOR*
> *END.*

5.5 Experiments and Simulation

In this section we discuss the experimental results taking a benchmark database described in Fig. 5.6.

5.5.1 Basic Assumptions

In this experiment we have considered the following assumptions

- The network has 21 hexagonal cells and 7 channels.
- The value of
  - Co-channel distance is 2
  - Adjacency channel distance is 3
  - Co-site distance is 4
- The number of incoming calls lies in the range 0 to 150 and changes dynamically.
- Initial population size was taken as 20.
- The velocity change was from 0 to 120 km/hr.
- The distance between two base stations is 2 km and remains unchanged.
- The calls are considered long if they go on more than 30 minutes.
- The direction of the MS, its distance from the base station and velocity changes dynamically.

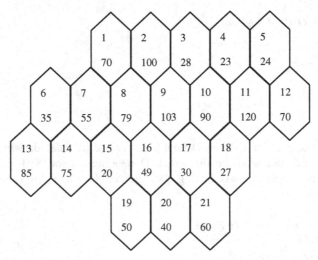

Fig. 5.6 Network of 21 cells

The algorithm stated above is executed using the given SVI. The initial conditions and necessary changes of the dynamic network are enforced obeying the above stated assumptions. The results obtained are shown an explained below.

5.5.2 Results

This is clear from Fig 5.7a , Fig 5.7b and Fig 5.7c that the problem of call admission control is optimized using BBO for the different ranges of velocities. It is seen that in low to moderate high velocity the algorithm using BBO works very well.

The graphical representation of the fitness values dependent on various hotness value range are given in the following figures Fig.5.8a, Fig 5.8b and Fig 5.8c. here we start with the assumption that the network is already in work and see that in all the cases the BBO algorithm is giving a very commendable result.

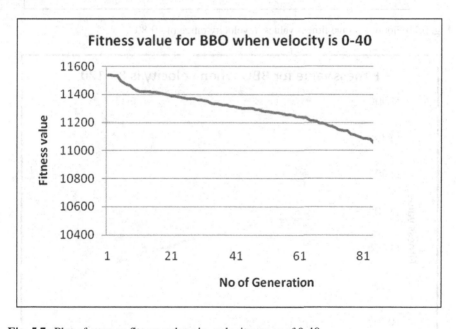

Fig. 5.7a Plot of average fitness values in velocity range of 0-40

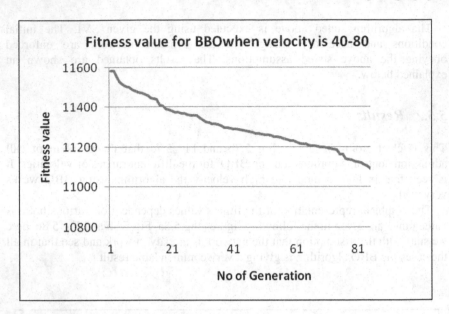

Fig. 5.7b Plot of average fitness values in velocity range of 40-80

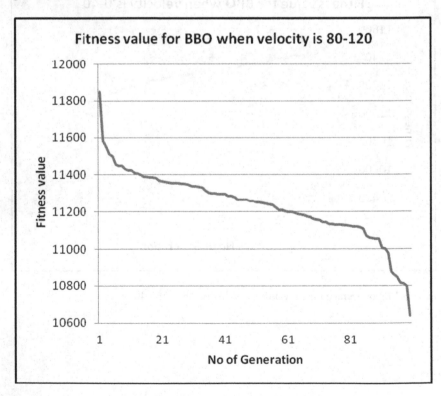

Fig. 5.7c Plot of average fitness values invelocity range of 80-120

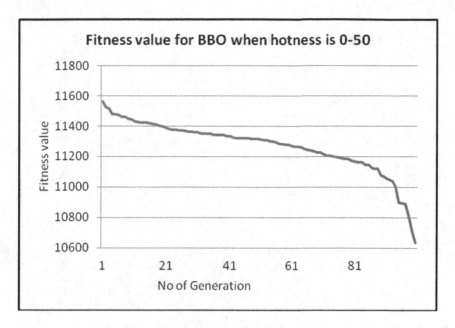

Fig. 5.8a Plot of average fitness values in hotness range of 0-50

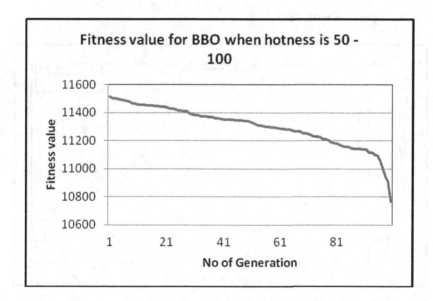

Fig. 5.8b Plot of average fitness values in hotness range of 50-100

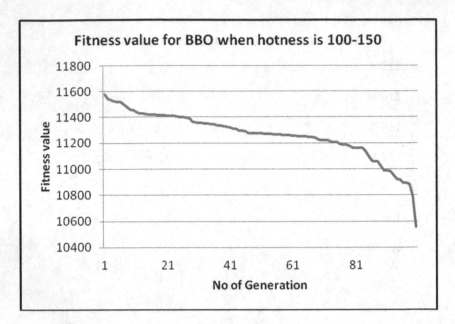

Fig. 5.8c Plot of average fitness values in hotness range of 100-150

5.5.3 Comparison with PSO

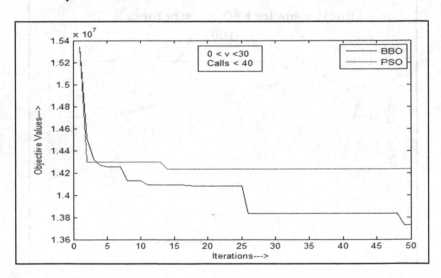

Fig. 5.9a Comparison using BBO and PSO in low hotness and velocity

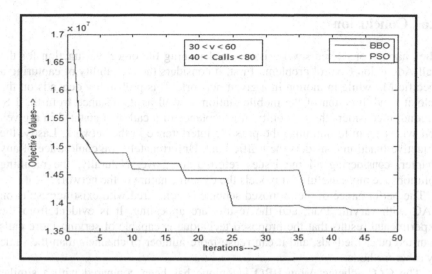

Fig 5.9b Comparison using BBO and PSO in moderate velocity and hotness

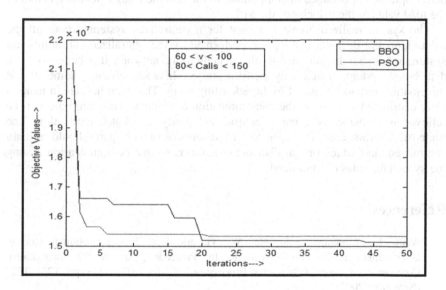

Fig. 5.9c Comparison using BBO and PSO in high velocity and hotness

It is evident from figure 5.9a, 5.9b and 5.9c that BBO works better than PSO. BBO may not reach the optimal solution faster but it gives a better result at the end.

5.6 Conclusion

The chapter considered several issues in designing the objective function for the Call Admission Control problem. First, it considers the probability of capturing a specific MS while in motion in a given network. This probability depends on the velocity and direction of the mobile station as well as its distance from the BS. Second, it considers the feasibility of acceptance of a call by a channel in a given cell with an aim to minimize the possible interference in the network. Lastly, the objective function considers the traffic load. Unfortunately, we could trace of any chapter, considering all the issues referred to above. Naturally, the resulting solutions are more useful as it models the dynamic nature of the network.

The performance of the proposed scheme is compared with existing results on CAC with varying load, and the results are appealing. It is evident from the experimental results that the proposed technique is capable of serving more calls than the other methods, and it can reuse more number of channels than the same by other methods.

The CAC scheme using BBO technique has been compared with a similar scheme using PSO. The results show that when BBO scheme is considered, a better optimum is obtained in comparison to the scheme using PSO irrespective of the load value or the velocity of the MS.

The system realized above is meant for a centralized system, where all the calculation and decision making are performed by the centralized call admission system and is then broadcasted to all the cells in the network. It is better to use a distributed system, which may perform faster. In such systems, some of the information should be shared by broadcasting them. The main hurdles in using a fully distributed system are the communication bottleneck. This can, however, be relieved by considering a partly central and partly distributed realization. The trade-off in this case is important to determine which part should remain centralized, and which one in distributed manner, so that communication among the system modules is minimized.

References

1. Wang, L., Arunkumaar, S., Gu, W.: Genetic algorithms for optimal channel assignment inmobile communications. In: Proceedings of the 9th International Conference on Neural Information Processing ICONIP 2002, vol. 3, pp. 1221–1225 (November 2002)
2. Hou, J., Fang, Y.: Mobility-based call admission control schemes for wireless mobile networks. Wirel. Commun. Mob. Comput. 1, 269–282 (2001)
3. Sakar, S., Sivarajan, K.N.: Channel assignment algorithms satisfyingm cochannel and adjacent channel reuse constraints in cellular mobile networks. IEEE Trans. Veh. Technol. 51(5), 954–967 (2002)
4. Wang, S.-L., Hou, Y.-B., Huang, J.-H., Huang, Z.-Q.: Adaptive Call Admission Control Based on Enhanced Genetic Algorithm in Wireless/Mobile Network. In: Proceedings of the 18th IEEE International Conference on Tools with Artificial Intelligence (2006)

5. Chen, H., Cheng, C.-C., Yeh, H.-H.: Guard-Channel-Based Incremental and Dynamic Optimization on Call Admission Control for Next-Generation QoS-Aware Heterogeneous Systems. IEEE Transaction on Vehicular Technology 57(5) (September 2008)

6. Thilakawardana, S., Tafazolli, R.: Efficient Call Admission Control and Scheduling Technique for GPRS Using Genetic Algorithms. Mobile Communications Research Group, Centre for Communications Systems Research, CCSR (2004)

7. Kang, S.H., Sung, D.K.: A CAC Scheme Based on Real-Time Cell Loss Estimation for ATM Multiplexers. IEEE Transaction on Communication 48(2) (February 2000)

8. Xiao, Y., Chen, C.L.P., Wang, Y.: A Near Optimal Call Admission Control With Genetic Algorithm For Multimedia Services In Wireless/Mobile Networks. In: IEEE Nat. Aero. Elec. Conf., pp. 787–792 (October 2000)

9. Deb, K.: Multi-Objective Optimization using Evolutionary Algorithms. John Wiley, London (2001)

10. Nuaymi, L., Godlewski, P., Mihailescu, C.: Call Admission Control Algorithm for Cellular CDMA Systems based on Best Achievable Performance. In: Proc. IEEE Vehic. Tech. Conf. (VT 2000 - Spring), Tokyo, vol. 1, pp. 375–379 (2000)

11. Jeon, W., Jeong, D.: Call Admission Control for Mobile Multimedia Communications with Traffic Asymmetry between Uplink and Downlink. IEEE Trans. Vehic. Tech. 50(1), 59–66 (2001)

12. Koo, I., Bahng, S., Kim, K.: Resource reservation in call admission control schemes for CDMA systems with nonuniform traffic distribution among cells. In: Proc. IEEE Veh. Technol. Conf., vol. 1, pp. 438–441 (2003)

13. Tugcu, T., Ersoy, C.: Application of a realistic mobility model to call admissions in DS-CDMA cellular systems. In: Proc. IEEE Veh. Technol. Conf., vol. 2, pp. 1047–1051 (2001)

14. Badia, L., Zorzi, M., Gazzini, A.: On the impact of user mobility on call admission control in WCDMA systems. In: Proc. IEEE Veh. Technol. Conf., vol. 1, pp. 121–126 (September 2002)

15. Ahmed, M.H.: Memorial Call Admission Control In Wireless Networks: A Comprehensive Survey. IEEE Communication Surveys 7(1) (2005)

16. Tian, X., Ji, C.: Bounding the Performance of Dynamic Channel Allocation with QoS Provisioning for Distributed Admission Control in Wireless Networks. IEEE Trans. Vehic. Tech. 50(2), 388–397 (2001)

17. Haung, Y., Ho, J.: Distributed Call Admission Control for a Heterogeneous PCS Network Computer. IEEE Trans. Comp. 5(12), 1400–1409 (2002)

18. Islam, M.M., Murshed, M., Dooley, L.S.: A directionally based bandwidth reservation scheme for call admission control. In: 5th International Conference on Computer and Information Technology (ICCIT 2002), Dhaka, Bangladesh, December 26-28 (2002)

19. Wong, D., Lim, T.J.: Soft handoffs in CDMA mobile systems. IEEE Personal Communications Magazine 4(6), 6–17 (1997)

20. Ye, J., Shen, X., Mark, J.W.: Call Admission Control in Wideband CDMA Cellular Networks by Using Fuzzy Logic. IEEE Transactions on Mobile Computing 4(2) (March/April 2005)

21. Chen, Y.-H., Chang, C.-J., Shen, S.: Outage-based fuzzy call admission controller with multi-user detection for WCDMA systems. IEE Proc. Commun. 152(5) (October 2005)

22. Chung-Ju, Chang, L.-C., Kuo, Y.-S., Chen, S.S.: Neural Fuzzy Call Admission and Rate Controller for WCDMA Cellular Systems Providing Multirate Services. In: IWCMC 2006, Vancouver, British Columbia, Canada, July 3-6 (2006)
23. Malarkkan, S., Ravichandran, V.C.: Performance analysis of call admission control in WCDMA System with Adaptive Multi Class Traffic based on Fuzzy Logic. IJCSNS International Journal of Computer Science and Network Security 6(11) (November 2006)
24. Yoshida, T., Watanabe, M., Nishida, S.: Channel prediction for OFDMA using mixtures of experts. International Journal of Knowledge-Based and Intelligent Engineering Systems 10(3), 193–200 (2006)
25. Moura, A.I., Ribeiro, C.H.C., Costa, A.H.R.: WBLS: A signal presence-based Wi-Fi localisation system for mobile devices in smart environments. International Journal of Knowledge-Based and Intelligent Engineering Systems 13(1), 5–18 (2009)
26. Liu, Z., Zarki, M.E.: SIR-Based Call Admission Control for DSCDMA Cellular System. IEEE J. Selected Areas of Comm. 12, 638–644 (1994)
27. Evans, J.S., Everitt, D.: Effective Bandwidth-Based Admission Control for Multiservice CDMA Cellular Networks. IEEE Trans. Vehicular Technology 48, 36–46 (1999)
28. Chang, C., Shen, S., Lin, J., Ren, F.: Intelligent Call Admission Control for Differentiated QoS Provisioning in Wide Band CDMA Cellular System. In: Proc. IEEE Vehicular Technology Conf. (VTC 2000), pp. 1057–1063 (2000)
29. Comaniciu, C., Mandayam, N., Famolari, D., Agrawal, P.: QoS Guarantees for Third Generation (3G) CDMA Systems Via Admission and Flow Control. In: Proc. IEEE Vehicular Technology Conf. (VTC 2000), pp. 249–256 (2000)
30. Bambos, N., Chen, S.C., Pottie, G.J.: Channel Access Algorithms with Active Link Protection for Wireless Communication Networks with Power Control. IEEE/ACM Trans. Networking 8, 583–597 (2000)
31. Andersin, M., Rosberg, Z., Zander, J.: Soft and Safe Admission Control in Cellular Networks. IEEE/ACM Trans. Networking 5, 255–265 (1997)
32. Ho, C.-J., Copeland, J.A., Lea, C.-T., Stüber, G.L.: On Call Admission Control in DS/CDMA Cellular Networks. IEEE Trans. Vehicular Technology 50, 1328–1343 (2001)
33. Fang, Y., Zhang, Y.: Call Admission Control Schemes and Performance Analysis in Wireless Mobile Networks. IEEE Trans. Vehicular Technology 51, 371–382 (2002)
34. Wu, S., Wong, K.Y.M., Li, B.: A Dynamic Call Admission Policy with Precision QoS Guarantee Using Stochastic Control for Mobile Wireless Networks. IEEE/ACM Trans. Networking 10, 257–271 (2002)
35. Fang, Y., Chlamtac, I.: Teletraffic Analysis and Mobility Modelling of PCS Networks. IEEE Trans. Comm. 47, 1062–1072 (1999)
36. Naghshineh, M., Schwartz, M.: Distributed Call Admission Control in Mobile/Wireless Networks. IEEE J. Select. Areas Comm. 14, 711–717 (1996)
37. Dziong, Z., Jia, M., Mermelstein, P.: Adaptive Traffic Admission for Integrated Services in CDMA Wireless-Access Networks. IEEE J. Seleced Areas of Comm. 14, 1737–1747 (1996)
38. Klir, G.J., Yuan, B.: Fuzzy Sets and Fuzzy Logic: Theory and Applications. Prentice-Hall (1995)
39. Wang, L.X.: A Course in Fuzzy Systems and Control. Prentice Hall (1997)
40. Rappaport, T.S.: Wireless Communications: Principles and Practice. Prentice Hall (1996)

41. Viterbi, A.J., Viterbi, A.M., Gilhousen, K.S., Zehavi, E.: Soft Handoff Extends CDMA Cell Coverage and Increase Reverse Link Capacity. IEEE J. Selected Areas of Comm. 12, 1281–1288 (1994)
42. Braae, M., Rutherford, D.A.: Fuzzy Relations in a Control Setting. Kybernetes 7, 185–188 (1978)
43. Del Re, E., Frantacci, R., Giambene, G.: Handover and Dynamic Channel Allocation Techniques in Mobile Cellular Networks. IEEE Trans. Vehicular Technology 44, 229–237 (1995)
44. Liu, T., Bahl, P., Chlamtac, I.: Mobility Modelling, Location Tracking, and Trajectory Prediction in Wireless ATM Networks. IEEE J. Selected Areas Comm. 16, 922–936 (1998)
45. Guerin, R.A.: Channel Occupancy Time Distribution in a Cellular Radio System. IEEE Trans. Vehicular Technology 36, 89–99 (1987)
46. Seskar, I., Maric, S., Holtzman, J., Wasserman, J.: Rate of Location Area Updates in Cellular Systems. In: Proc. IEEE Vehicular Technology Conc. (VTC 1992), pp. 694–697 (1992)
47. Nanda, S.: Teletraffic Models for Urban and Suburban Microcells: Cell Sizes and Handoff Rates. IEEE Trans. Vehicular Technology 42, 673–682 (1993)
48. Varshney, U., Jain, R.: Issues in emerging 4G wireless networks. IEEE Computer 34(6), 94–96 (2001)
49. Simon, D.: Biogeography-based optimization. IEEE Transactions on Evolutionary Computation 12, 702–713 (2008)
50. Simon, D., Ergezer, M., Du, D.: Markov analysis of biogeography-based optimization, http://academic.csuohio.edu/simond/bbo/markov/
51. Simon, D., Ergezer, M., Du, D.: Population distributions in biogeography-based optimization with elitism, http://academic.csuohio.edu/simond/bbo/markov/
52. Simon, D.: A probabilistic analysis of a simplified biogeography-based optimization algorithm, http://academic.csuohio.edu/simond/bbo/simplified/

Chapter 6
Conclusions and Future Directions

6.1 Conclusions

The thesis examined different approaches for call admission in a mobile cellular network. Existing models on call admission control usually ignores movement of the mobile station, and thus in most circumstances the allocation is done in a relatively *ad hoc* manner. The problem becomes more prominent particularly, when the mobile station crosses cell boundaries. In the present scenario, the direction of motion of the mobile station is predicted, and consequently the decisions about channel allocation for the new current call in progress can be taken up ahead of time.

The merit of the thesis lies in handling the call admission and dynamic channel assignment problem in a composite sense, which has not been taken up in any previous literature. Furthermore, the strategies used for call admission, including feasibility, hotness and availability criteria, together are different with respect to classical approaches adopted for the same problem.

The thesis proposed three different techniques for call admission control. The first approach deals with fuzzy condition sensitive rule firing to derive fuzzy inferences. A list of three parameters including feasibility, hotness and availability, distance from the base station, and speed of the mobile station is employed to test the firing conditions of the rules in order to derive fuzzy inferences. Computer simulation reveals that the proposed scheme reduces call drop as the assignment satisfies both the pre-conditions for feasibility and availability.

While the first approach refers to cell-wise call admission, the latter approaches consider the problem for the entire network compositely. This is realized by formulating call admission control as an optimization problem. The objective function used here attempts to minimize the call drops by jointly improving feasibility and probability of capture of mobile stations and reducing network loads. Two popular optimization techniques have been employed here to minimize the call drops. The techniques used are popularly known as Genetic Algorithm (GA) and Biogeography Based optimization (BBO). The results are compared with other well known techniques, including Differential Evolution and Particle Swarm Optimization. Experimental results indicate that BBO outperforms other techniques with respect to better accuracy in call assignment and soft handoff.

S. Ghosh and A. Konar: CAC in a Mobile Cellular Network, SCI 437, pp. 183–184.
springerlink.com   © Springer-Verlag Berlin Heidelberg 2013

6.2 Possible Direction of Future Research

The CAC problem addressed in the thesis has been solved by three interesting techniques of computational intelligence. However, there exist several hundreds techniques, falling within and outside the domain of computational intelligence, which can directly or indirectly with some modifications be applied to handle the CAC problems. The whole study of the CAC problem and its solution by all possible techniques, and comparison of their relative merits/demerits still remains an unsolved problem. Further, the thesis considers only fewer problems associated with the CAC problem. For example, the classical approaches to CAC problem ignore movement of the mobile stations, which has been taken care in this thesis. However, there are many other issues too, which could not be taken up in the present formulation of the CAC problem. Some of these issues, which are of practical importance in connection with the CAC problem are power optimization, consideration of call assignment delay, and call transfer time required during soft handoff.

Besides the above issue, we here considered a central administration system for call assignment. The central system usually has a manager which takes care of call allocation in the entire network irrespective of call demands in individual cells. The central system is relatively slower as it has plenty of input calls and thus optimization in call management takes a lot of time for the inherent complexity of the problem. To alleviate the drawbacks of the central systems, distributed systems are currently gaining importance, where instead of having a central manager; each cell has its own manager to take care of call admission in a cell. Although managers of the cells can function almost independently, they need to take care of call transfers from other cells to itself and vice-versa. Besides the above, the call assignment undertaken by a cell needs to be communicated to the nearest neighbours, so that these neighbours do not assign adjacent channels to pending/incoming calls in those cells.

In recent times, researchers are taking interest to consider hybridization of the above two call assignment policies to derive benefit from both of these without sacrificing merits of individual approaches. In a hybrid CAC system, inter-cell communication is accomplished through central CAC system, whereas the decisions about call admission are undertaken by the cell itself. The decision about soft handoff is taken by the central management system, while decision about assignment of new calls and pending calls are undertaken by the local call admission controllers.

Appendix A: *Program Realization of CAC Using Fuzzy Threshold Logic*

Here the program for Call Admission Control in Mobile network is presented in detail. The program has been written in C with standard header files for input output and library functions. Data stored in files and are used as input given as

c.txt	This file is used for saving the compatibility matrix.
h.txt	This file is saving the number of incoming call at any instance.
nb.txt	In this file the cell organization of the network is stored
f.txt	Here the initial channel allocation matrix is stored
stat.txt	The final matrix of allocation is stored

FUZZY TO BINARY CAC

The program starts with a main function whose return type is void. All the input files are opened in "w+" mode for both reading and updating.

Step 1.

In this part of the program, the network is defined with every cell position with respect to its neighbor.

```
for(i=0;i<21;i++){
        printf("Neighbour of cell %d are : ",i+1);
        for(j=0;j<6;j++){
        fscanf(nf,"%d",&nb[i][j]);
        printf(" %d ",nb[i][j]);
        }    printf("\n");}
        printf("\n\n");
```

Step 2.

In this part the compatibility matrix is loaded for further use in feasibility checking.

```
printf("Compatibility matrix fir 21 cels with 7 channel is given by :\n\n");
for(i=0;i<21;i++){
        for(j=0;j<21;j++){
        fscanf(cf,"%d",&c[i][j]);
        printf("%d ",c[i][j]);
        }    printf("\n");}

        printf("\n\n\n\n");
for(i=0;i<21;i++){
```

Step 3.
In this section, the availability matrix is constructed using current allocation matrix

```
printf("free channel in cell   %d are: ",i+1);
        for(j=0;j<7;j++){
            fscanf(ff,"%d",&f[i][j]);
            avl[i][j]=1-f[i][j];
            printf("%d ",avl[i][j]);
        }  printf("\n");}
                printf("\n\n");
```

Step 4.
This this step the demand matrix is lodaded for further use.

```
printf("demand in all 21 cells are: \n\n");
for(i=0;i<21;i++){
 fscanf(hf,"%d",&h[i]);
 printf("%d ",h[i]);
        }
```

Step 5.
In this part the velocity of MS and the angle of their direction is loaded .

```
randomize();
vel=random(100);
th=random(3);
printf("\n\n\nvel=%d, thresh= %d",vel,th);
```

Step 6.
From here fuzzy to binary decision making starts. The iterative loop starts for calculating the feasibility of a call to be assigned to a channel in a cell.

```
Ita=0;
for(y=0;y<50;y++){  if(ita==200) exit(status-'0');
  for(r=0;h[r]>0;r++,ita++){
            dir=random(6);
            printf("\n\n\n dir=%d",dir+1);
            l=large();
            printf("\n%d th cell have largest dimand\n" ,l+1);
            printf("%d",l);
            printf("\n neigbour of this cell are: ");
            for(j=0;j<6;j++){
                    n[j]=nb[l][j];
                     printf("%d ",n[j]);}
                    k=n[dir];
                    if(k==0)continue;
                    printf("\n in %d th nbd the available channels are : ",k);
```

```
for(i=0;i<7;i++){
if(avl[k-1][i]==1)
printf("%d ",i+1);                                          }
printf("\n");
printf("\nvalue of compatibility matrix is %d\n",c[k][l]);
```

Step 7.
Decision making process starts.

```
switch(c[k][l]){
```

Step 8: DECISION FOR 2nd NEAR NEIGHBOUR

Here we take decision about the channels who are engaged in   2nd nearest neighbour*/

```
case(3): {for(x=0;x<7;x++){
        if(avl[k-1][x]==0)
        p=x;
        printf("\n1st busy channel is %d \n",x);
        break;}
        for(x=0;x<7;x++){
        f(avl[k-1][x]==0)
        a[x]=-1;
        else a[x]=x+1 ;
        printf("\n avl=%d ,a[%d]= %d \n",avl[k-1][x],x,x+1);        }
                for(x=0;x<7;x++){
                cdiv=a[x]-p; printf("\ncdiv= %d \n",cdiv);
                if(cdiv>=3){
                avl[k-1][x]=0;
                break;}
                }       break;}
```

Step9: DECISION FOR NEXT NEIGHBOUR

Here we take decision about the channels who are engaged in   next nearest neighbour*/

```
        case(2):{for(x=0;x<7;x++){
        if(avl[k-1][x]==0)
        p=x;printf("\n1st busy channel is %d \n",x);
                break;}
                for(x=0;x<7;x++){
                if(avl[k-1][x]==0)
                a[x]=-1;
                else a[x]=x+1;
                printf("\n avl=%d ,a[%d]= %d \n",avl[k-1][x],x,x+1); }
        for(x=0;x<7;x++){
                cdiv=a[x]-p; printf("\ncdiv= %d \n",cdiv);
```

```
                if(cdiv>=2){
                avl[k-1][x]=0;
                break;}
                        }break;  }
```

Step10:   DECISION TAKEN INSIDE THE CELL
Here we take decision about the channels who are engaged inside the cell */

```
        case(0): {for(x=0;x<7;x++){
                if(avl[k-1][x]==1)
                avl[k-1][x]=0;break;}break;}
```

Step 11:   DECISION FOR NO INTERFERENCE
Here we take decision about the channels who are engaged beyond 2nd nearest neighbour. In those case it is considered that there is no interference */

```
case(5):{for(x=0;x<7;x++){
                if(avl[l][x]==0)
                p=x+1;printf("\n1st assign channel is %d \n",p);
                break;}
                for(x=0;x<7;x++){
                if(avl[l][x]==0)
                a[x]=-1;
                else a[x]=x+1;
                printf("\n avl=%d ,a[%d]= %d \n",avl[k-1][x],x,x+1);              }
        for(x=0;x<7;x++){
                cdiv=a[x]-p; printf("\ncdiv= %d \n",cdiv);
                if(cdiv>=5){
                avl[l][x]=0;
                break;}
                        }break;  } }
```

Step 12: ASSIGNMENT OF CHANNEL
Here the calls are assigned to channels*/

```
if(avl[k-1][x]==1)dr++;
        else asn++;
                h[l]--;
                randomize();
                hup=random(21);
                hfreq=random(7);
                nucal=random(21);
                avl[hup][hfreq]=1;
```

Step 13: REFRESHING HOTNESS TABLE

```
                    h[nucal]=h[nucal]+1;
                    printf("\nnew call %d hangup: %d ",nucal,avl[hup][hfreq]);
                    printf("\n new dimand %d",h[nucal] );
                    printf("\nnew demand in all 21 cells are: \n\n");
      for(i=0;i<21;i++){
         printf("%d ",h[i]);
              }
                fflush(stdin);
              printf(" \n press any key for next ...................");
              ch=getchar();
                         for(i=0;i<21;i++){
```

Step14: REFRESHING ALLOCATION TABLE

```
printf("free channel in cell   %d are:            ",i+1);
          for(j=0;j<7;j++){
              printf("%d ",avl[i][j]);
            }   printf("\n");}
        printf(" \nno of itaration = %d \nno of drop= %d no of
assign=%d",ita+1,dr,asn);
                         fprintf(fstat, " %d, %d,   %d\n", ita+1,dr,asn);
                         printf("\n\n");
                                     } }
int large(void){
 int i,j=0;
 for(i=1;i<21;i++)
{
 if(h[i]>h[j])
  j=i;
}
   return(j);
 }
```

FUZZY TO FUZZY CALL ADMISSION CONTROL

In this program same input files are required .

/* **FIXING THE NETWORK**
In this part of the program the network is defined with every cell position with respect to its neighbour.*/

```
for(i=0;i<21;i++){
        printf("Neighbour of cell %d are : ",i+1);
        for(j=0;j<6;j++){
         fscanf(nf,"%d",&nb[i][j]);
         printf(" %d ",nb[i][j]);
         }     printf("\n");}
        printf("\n\n");
```

ita=0;

/* **TAKING THE COMPATIBILITY MATRIX**
In this part the compatibility matrix is loaded */

```
printf("Compatibility matrix fir 21 cels with 7 channel is given by :\n\n");
for(i=0;i<21;i++){
        for(j=0;j<21;j++){
        fscanf(cf,"%d",&c[i][j]);
        printf("%d ",c[i][j]);
        }     printf("\n");}

        printf("\n\n\n\n");
for(i=0;i<21;i++){
```

/* **GENERATING THE AVAILABILITY MATRIX**
In this section the availability matrix is constructed using current allocation matrix*/

```
printf("free channel in cell   %d are:            ",i+1);
        for(j=0;j<7;j++){
        fscanf(ff,"%d",&f[i][j]);
        avl[i][j]=1-f[i][j];
        printf("%d ",avl[i][j]);
        }   printf("\n");}
                printf("\n\n");
```

/* **TAKING THE DEMAND**
Demand is lodaded */
printf("demand in all 21 cells are: \n\n");

```
for(i=0;i<21;i++){
 fscanf(hf,"%d",&h[i]);
 printf("%d ",h[i]);
          }
```

/ GETTING VELOCITY AND ANGLE FOR MS
In tjis part the velocity of MS and the angle of their direction is loaded*/

```
 randomize();
 vel=random(100);
 th=random(3);
 printf("\n\n\nvel=%d, thresh= %d",vel,th);
```

/* START THE FUZZY BINARY DECISION MAKING
From here fuzzy to binary decision making starts */

```
for(y=0;y<50;y++){ if(ita==200) exit(status-'0');
 for(r=0;h[r]>0;r++,ita++){
          dir=random(6);
          printf("\n\n\n dir=%d",dir+1);
          l=large();
          printf("\n%d th cell have largest dimand\n" ,l+1);
          printf("%d",l);
          printf("\n neigbour of this cell are: ");
          for(j=0;j<6;j++){
                  n[j]=nb[l][j];
                   printf("%d ",n[j]);}
                  k=n[dir];
                  if(k==0)continue;
                  printf("\n in %d th nbd the available channels are : ",k);
                  for(i=0;i<7;i++){
                  if(avl[k-1][i]==1)
                  printf("%d ",i+1);                                    }
                  printf("\n");
                  printf("\nvalue of compatibility matrix is %d\n",c[k][l]);
```

// DECISION MAKING
```
switch(c[k][l]){
```

/ *DECISION FOR 2nd NEAREST NEIGHBOUR
Here we take decision about the channels who are engaged in 2nd nearest
neighbour*/

```
case(3): {for(x=0;x<7;x++){
          if(avl[k-1][x]==0)
            p=x;
```

```
    printf("\n1st busy channel is %d \n",x);
    break;}
   for(x=0;x<7;x++){
f(avl[k-1][x]==0)
a[x]=-1;
else a[x]=x+1 ;
   printf("\n avl=%d ,a[%d]= %d \n",avl[k-1][x],x,x+1);          }
           for(x=0;x<7;x++){
           cdiv=a[x]-p; printf("\ncdiv= %d \n",cdiv);
           if(cdiv>=3){
           avl[k-1][x]=0;
           break;}
           }          break;}
```

/* **DECISION FOR NEXT NEIGHBOUR**
Here we take decision about the channels who are engaged in   next nearest neighbour*/

```
    case(2):{for(x=0;x<7;x++){
   if(avl[k-1][x]==0)
     p=x;printf("\n1st busy channel is %d \n",x);
             break;}
           for(x=0;x<7;x++){
           if(avl[k-1][x]==0)
           a[x]=-1;
           else a[x]=x+1;
           printf("\n avl=%d ,a[%d]= %d \n",avl[k-1][x],x,x+1); }
   for(x=0;x<7;x++){
           cdiv=a[x]-p; printf("\ncdiv= %d \n",cdiv);
           if(cdiv>=2){
           avl[k-1][x]=0;
           break;}
                   }break;  }
```

/* **DECISION TAKEN INSIDE THE CELL**
Here we take decision about the channels who are engaged inside the cell */

```
    case(0): {for(x=0;x<7;x++){
```

$$\mu_{SHO}(uc) = min\,[1, max\,\{\mu_{MID}(avl_i), \mu_{LO}(avl_i)\}^* \max_{j}\{\mu_{LO}(feas^j)\} +$$

$$\max_{N}\{\mu_{LO}(hot_N)\}^* \max_{N}\{\mu_{HI}(avl_N)\}].$$

```
}
```
IF $\mu_{SHO} < \varepsilon$

```
           avl[k-1][x]=0;break;}break;}
```

/* DECISION FOR NO INTERFERENCE

Here we take decision about the channels who are engaged beyond 2nd nearest neighbour. In those case it is considered that there is no interference */

case(5):{for(x=0;x<7;x++){

$$\mu_{SHO}\ (uc) = min\ [1, max\ \{\mu_{MID}\ (avl_{\ i}\ ), \mu_{LO}\ (avl_{\ i}\ )\}^* \max_{j}\ \{\mu_{LO}\ (feas^{\ j}\ )\} +$$

$$\max_{N}\ \{\mu_{LO}\ (hot_{\ N}\ )\}^* \max_{N}\ \{\mu_{HI}\ (avl_{\ N}\ )\}\ ].$$

```
        }
    IF   μSHO < ε
                    p=x+1;                  break;}}
                    for(x=0;x<7;x++){
                    if(avl[l][x]==0)
                    a[x]=-1;
                    else a[x]=x+1;
                     printf("\n avl=%d ,a[%d]= %d \n",avl[k-1][x],x,x+1);          }
              for(x=0;x<7;x++){
                         cdiv=a[x]-p; printf("\ncdiv= %d \n",cdiv);
                         if(cdiv>=5){
                         avl[l][x]=0;
                         break;}
                                }break;         } }
```

/* ASSIGNMENT OF CHANNEL

Here the calls are assigned to channels*/

```
    if(avl[k-1][x]==1)dr++;
              else asn++;
                       h[l]--;
                    randomize();
                     hup=random(21);
                     hfreq=random(7);
                     nucal=random(21);
                     avl[hup][hfreq]=1;
```

// REFRESHING HOTNESS TABLE

```
                    h[nucal]=h[nucal]+1;
                    printf("\nnew call %d hangup: %d ",nucal,avl[hup][hfreq]);
                    printf("\n new dimand %d",h[nucal] );
                    printf("\nnew demand in all 21 cells are: \n\n");
    for(i=0;i<21;i++){
      printf("%d ",h[i]);
                }
```

```
            fflush(stdin);
            printf(" \n press any key for next ..................");
            ch=getchar();
                        for(i=0;i<21;i++){
```

// REFRESHING ALLOCATION TABLE

```
printf("free channel in cell   %d are:          ",i+1);
        for(j=0;j<7;j++){
          printf("%d ",avl[i][j]);
          }   printf("\n");}
      printf(" \nno of itaration = %d \nno of drop= %d no of
asign=%d",ita+1,dr,asn);
                    fprintf(fstat, " %d, %d,   %d\n", ita+1,dr,asn);
                      printf("\n\n");
                                } }
int large(void){
 int i,j=0;
 for(i=1;i<21;i++)
{
 if(h[i]>h[j])
  j=i;
}
  return(j);
 }
```

Appendix B: *Program Realization of CAC Using Genetic Algorithm*

```c
#include<stdio.h>
#include<stdlib.h>
#include<time.h>
#include<math.h>
#define PI 3.14159

int vel[21][7],c[21][21],h[21],nb[21][6],avl[21][7],sum;
int P[20][21][7],par1[21][7],par2[21][7],Cn1[21][7],Cn2[21][7],hn[21];
float dis[21][7],Feas[20],T[21][7];
 char ch;
void main(){

                int i,j,k,a,b,Ite,row,col,x1,y1,x2,y2;
                float fitp1,fitp2,fitc1,fitc2,temp;
                float fit(int a);
                FILE *cf,*ff,*hf, *nf,*fr,*of,*s,*h1,*f1,*f2;
                cf=fopen("c.txt","r");
                ff=fopen("per.txt","w+")
                hf=fopen("h.txt","a+");
                nf=fopen("nb.txt","r");
                fr=fopen("TV.txt","w+");
                of=fopen("newjen.txt","w+");
                s=fopen("s.txt","w+");
                h1=fopen("h1.txt","w+");
                f1=fopen("fit1.txt","w+");
                f2=fopen("fit2.txt","w+");
                printf("The 6 neighbour of each cell are as follows :\n\n");
                for(i=0;i<21;i++){
                        printf("Neighbour of cell %d are : ",i+1);
                        for(j=0;j<6;j++){
                          fscanf(nf,"%d",&nb[i][j]);
                                printf(" %d ",nb[i][j]);
                                }printf("\n");
                                }
        printf("\n\n");
/* LOADING THE COMPATIBILITY MATRIX

        printf("Compatibility matrix fir 21 cels with 7 channel is given by :\n\n");
                for(i=0;i<21;i++){
                        for(j=0;j<21;j++){
```

```
                        fscanf(cf,"%d",&c[i][j]);
                        printf("%d ",c[i][j]);
                }printf("\n");
                        printf("\n\n\n\n");
```

/*LOADING THE HOTNESS, DEMAND, VELOCITY & ANGLE*/

```
printf("demand in all 21 cells are: \n\n");
                for(i=0;i<21;i++){
                                        fscanf(hf,"%d",&h[i]);
                                        printf("%d ",h[i]); }
                fputs("\n Velocity-\n ",fr);
                printf("\n\n velocity of the calls in MS: \n");
                randomize();
                for(j=0;j<21;j++){
                        for(k=0;k<7;k++){
                                        vel[j][k] = random(120);
                                         printf("%d        ",vel[j][k]);
                                        fprintf(fr,"%d        ",vel[j][k]);
fputs("\n ",fr);
                                                } fputs("\n ",fr);
                                        printf("\n");
                                                }
                        printf("\n\n angle of the calls in MS: \n");
                fputs("\n Angle:\n ",fr);
                randomize();
                for(j=0;j<21;j++){
                for(k=0;k<7;k++){
                                        T[j][k] = random(360)/ PI;
                                        fprintf(fr,"%f        ",T[j][k]);
                                        printf("%f   ",T[j][k]); fputs("\n ",fr);
                                } fputs("\n ",fr);
                        printf("\n");
                                                }
                        printf("\n\n Dis of the   MS from BS: \n");
                randomize();
                for(j=0;j<21;j++){
                for(k=0;k<7;k++){
                                        dis[j][k] = rand()/20000.00;
                                        printf("%f ",dis[j][k]);
                                                }
                                        printf("\n");

}
                fflush(stdin);
                printf("\n end of input press any key ..............\n");
```

```
                              ch=getchar();
                              printf("\n\n Random generation of parents\n");
                              randomize();
                              for(i=0;i<20;i++){
printf(" the %d th parent \n:",i);
fprintf(ff,"parent %d :\n\n",i);
          for(j=0;j<21;j++){
          for(k=0;k<7;k++){
          P[i][j][k]= random(2);
          printf("%d   ",P[i][j][k]);
          fprintf(ff,"%d",P[i][j][k]);
                    }
  fputs("\n ",ff);

printf("\n");

                    }fputs("\n ",ff);}
```

/* START ITERATION   */

```
          fflush(stdin);
          printf("\n starting itaration press any key ..............\n");
          ch=getchar();
          randomize();
          for(Ite=0;Ite<100;Ite++){
          printf("demand in all 21 cells are: \n\n");
          for(i=0;i<21;i++){
             hn[i]=random(150);
fprintf(h1,"%d \n",hn[i]);}
```

/* Random parent Generation */

```
for(a=0;a<20;a++){
b=random(20);
if(a==b){
printf("\n Both are same;");
continue;}
printf("\nthe 1st parent chosen is=%d and the second is=%d",a,b);
row=random(21);
col=random(7);
sum=0;
printf("\n The first parent:\n");
for (j=0;j<21;j++) {
for(k=0;k<7;k++){
par1[j][k]= P[a][j][k];
```

```
printf("%d ",par1[j][k]);
fprintf(of,"%d ",P[a][j][k]);
sum=sum+P[a][j][k];                    }
fputs("\n",of);
printf(s,"%d \t",sum);
fputs("\n ",s);
fitp1= fit(sum);
 printf("\nfitness of 1st parent=%f \n",fitp1);
 printf("\n"); }
sum=0;
printf("\n The second parent:\n");
for (j=0;j<21;j++) {
for(k=0;k<7;k++){
par2[j][k]=  P[b][j][k];
printf("%d ",par1[j][k]);
printf(of,"%d ",P[b][j][k]);
sum=sum+P[b][j][k];                    }
fputs("\n",of);
fprintf(s,"%d \t",sum);
fputs("\n ",s);
fitp2= fit(sum);
printf("\nfitness of 1st parent=%f \n",fitp2);
printf("\n"); }
 fprintf(of,"%d th generation\n",Ite);
```

/*CROSS OVER*/

```
for(j=0;j<row;j++)
           for(k=0;k<7;k++){
                   Cn1[0][k]=par2[j][k];
                   Cn2[0][k]=par1[j][k];
                   }
for(j=row;j<21;j++){
           for(k=0;k<col;k++){
           Cn1[j][0]=par2[j][k];
           Cn2[j][0]=par1[j][k];}
           for(k=col;k<7;k++){
           Cn1[j][k]=par2[j][k];
           Cn2[j][k]=par1[j][k];
           }   }
printf("\n The first child:\n")
sum=0;
for (j=0;j<21;j++) {
for(k=0;k<7;k++){
printf("%d ",Cn1[j][k]);
fprintf(of,"%d ",Cn1[j][k]);
```

```
sum=sum+Cn1[j][k];              }
fputs("\n",of);
fprintf(s,"%d \t",sum);
fputs("\n ",s);
fitc1= fit(sum);
printf("\nfitness of 1st parent=%f \n",fitc1);
printf("\n"); }


                    printf("\n The second child:\n");
                    sum=0;
                    for (j=0;j<21;j++) {
        for(k=0;k<7;k++){
        printf("%d ",Cn2[j][k]);
        fprintf(of,"%d ",Cn2[j][k]);
        sum=sum+Cn2[j][k];              }
        fputs("\n",of);
        fprintf(s,"%d \t",sum);
        fputs("\n ",s);
        fitc2= fit(sum);
        printf("fitness of 2nd child=%f",fitc2) ;
        printf("\n"); }
        if(fitp1>fitc1){
        fprintf(f1,"%f \n   ",fitc1);
        for(j=0,k=0;j<21,k<7;j++,k++)
                    P[a][j][k]=Cn1[j][k];
                        }
                            if(fitp2>fitc2){
                                fprintf(f2,"%f \n",fitc2);
                                for(j=0,k=0;j<21,k<7;j++,k++)
                                        P[b][j][k]=Cn2[j][k];
                                                            }


                * MUTATION*/

randomize();
for(j=0;j<21;j++){
        x1=random(21);
        y1=random(7);
        P[a][x1][y1]=0;
        x2=random(21);
        y2=random(7);
        P[b][x2][y2]=0; }
        printf("x1=%d,y1=%d,x1=%d,y2=%d",x1,y1,x2,y2);
   temp=sqrt( pow(fitp1-fitc1,2)+pow(fitp2-fitc2,2)   );
```

```
            printf("\ntemp= %f\n",temp);
            if(temp<.00001)
            break;
}}
}



float fit(int p){
int   ECS,EAC,E,j,k,m,n,a[21][7],ht,i;
float tp[21],tf,f,lod;

ECS=0; EAC=0;
```

/* FITNESS CALCULATION*/

```
/*/ fflush(stdin);
printf(" for G press any key ..............\n");
ch=getchar(); */
switch(p){
case (0):{
for(j=0;j<20;j++)
for(k=0;k<7;k++)
a[j][k]=par1[j][k];
break;}
case (1):{
for(j=0;j<20;j++)
for(k=0;k<7;k++)
a[j][k]=par2[j][k];
break;}
case (2):{
for(j=0;j<20;j++)
for(k=0;k<7;k++)
a[j][k]=Cn1[j][k];
break;}
case (3):{
for(j=0;j<20;j++)
for(k=0;k<7;k++)
a[j][k]=Cn2[j][k];
break;}
}

for(j=0;j<20;j++){
for(k=0;k<7;k++)
{       if(abs(a[j][k]-a[j][k+1])<c[j][j])
```

```
            ECS=ECS ++ ; }
            printf("\n");


}
for(j=0;j<21;j++)
 for(k=0;k<7;k++)
 for(m=0;m<21;m++)
 for(n=0;n<7;n++) {
 if(abs(a[j][k]-a[m][n])<c[j][m])
  EAC=EAC++;}
  E= ECS+EAC;
for(j=0;j<21;j++)tp[j]=0.0;tf=0.0;
 for(j=0;j<21;j++){
            for(k=0;k<7;k++){
                        tp[j]=tp[j]+(vel[j][k]);
                        // * cos(T[j][k]))*/
            } tf=tf + tp[j]/h[j];
              }
            ht=0;
            for(i=0;i<21;i++)
                             ht=ht+hn[i] ;
                             lod=p/ht ;
  f=(E+ tf - lod);

/* printf("\nESC=%d,EAC=%d, E=%d,tf=%f, t=%f\n",ECS,EAC,E,tf,f);    */
 return(f);


}
```

Appendix C: *Program Realization of CAC Using BBO*

>> ACO
This command would run ACO on the Step function (which is codified in Step.m).

```
function [MinCost] = ACO(ProblemFunction, DisplayFlag)

% Ant colony optimization algorithm for optimizing a general function.

% INPUTS: ProblemFunction is the handle of the function that returns
%             the handles of the initialization, cost, and feasibility functions.
%             DisplayFlag says whether or not to display information during
iterations and plot results.

if ~exist('DisplayFlag', 'var')
    DisplayFlag = true;
end

[OPTIONS, MinCost, AvgCost, InitFunction, CostFunction, FeasibleFunction, ...
    MaxParValue, MinParValue, Population] = Init(DisplayFlag,
ProblemFunction);

Keep = 2; % elitism parameter: how many of the best individuals to keep from one
generation to the next

% ACO parameter initialization
tau0 = 1e-6; % initial pheromone value, between 0 and 0.5
Q = 20; % pheromonone update constant, between 0 and 100
q0 = 1; % exploration constant, between 0 and 1
rhog = 0.9; % global pheromone decay rate, between 0 and 1
rhol = 0.5; % local pheromone decay rate, between 0 and 1
alpha = 1; % pheromone sensitivity, between 1 and 5
beta = 5; % visibility sensitivity, between 0 and 15
tau = tau0 * ones(MaxParValue-MinParValue+1, 1); % initial pheromone values
p = zeros(size(tau)); % allocate array for probabilities

% Begin the optimization loop
for GenIndex = 1 : OPTIONS.Maxgen
    % pheromone decay
    tau = (1 - rhog) * tau;
    % Use each solution to update the pheromone for each parameter value
```

```
for k = 1 : OPTIONS.popsize
    Cost = Population(k).cost;
    Chrom = Population(k).chrom;
    for i = 1 : length(Chrom)
        j = Chrom(i);
        if (Cost == 0)
            tau(j-MinParValue+1) = max(tau);
        else
            tau(j-MinParValue+1) = tau(j-MinParValue+1) + Q / Cost;
        end
    end
end
% Use the probabilities to generate new solutions
for k = Keep+1 : OPTIONS.popsize
    for j = 1 : OPTIONS.numVar
        % Generate probabilities based on pheromone amounts
        p = tau .^ alpha;
        p = p / sum(p);
        [Maxp, Maxpindex] = max(p);
        if rand < q0
            Select_index = Maxpindex;
        else
            SelectProb = p(1);
            Select_index = 1;
            RandomNumber = rand;
            while SelectProb < RandomNumber
                Select_index = Select_index + 1;
                if Select_index >= MaxParValue - MinParValue + 1
                    break;
                end
                SelectProb = SelectProb + p(Select_index);
            end
        end
        Population(k).chrom(j) = MinParValue + Select_index - 1;
        % local pheromone update
        tau(Select_index) = (1 - rhol) * tau(Select_index) + rhol * tau0;
    end
end
% Make sure the population does not have duplicates.
Population = ClearDups(Population, MaxParValue, MinParValue);
% Make sure each individual is legal.
Population = FeasibleFunction(OPTIONS, Population);
% Calculate cost
Population = CostFunction(OPTIONS, Population);
% Sort from best to worst
Population = PopSort(Population);
```

```
% Compute the average cost of the valid individuals
[AverageCost, nLegal] = ComputeAveCost(Population);
% Display info to screen
MinCost = [MinCost Population(1).cost];
AvgCost = [AvgCost AverageCost];
if DisplayFlag
    disp(['The best and mean of Generation # ', num2str(GenIndex), ' are ',...
        num2str(MinCost(end)), ' and ', num2str(AvgCost(end))]);
end
end
Conclude(DisplayFlag, OPTIONS, Population, nLegal, MinCost);
return;
```

Init - This contains various initialization settings for the optimization methods. You can edit this file to change the population size, the generation count limit, the problem dimension, and the mutation probability of any of the optimization methods that you want to run.

```
function [OPTIONS, MinCost, AvgCost, InitFunction, CostFunction,
FeasibleFunction, ...
    MaxParValue, MinParValue, Population] = Init(DisplayFlag, ProblemFunction,
RandSeed)

% Initialize population-based optimization software.

% WARNING: some of the optimization routines will not work if population size
is odd.
OPTIONS.popsize = 50; % total population size
OPTIONS.Maxgen = 50; % generation count limit
OPTIONS.numVar = 20; % number of genes in each population member
OPTIONS.pmutate = 0; % mutation probability

if ~exist('RandSeed', 'var')
    RandSeed = round(sum(100*clock));
end
rand('state', RandSeed); % initialize random number generator
if DisplayFlag
    disp(['random # seed = ', num2str(RandSeed)]);
end

% Get the addresses of the initialization, cost, and feasibility functions.
[InitFunction, CostFunction, FeasibleFunction] = ProblemFunction();
% Initialize the population.
[MaxParValue, MinParValue, Population, OPTIONS] = InitFunction(OPTIONS);
% Make sure the population does not have duplicates.
Population = ClearDups(Population, MaxParValue, MinParValue);
```

```
% Compute cost of each individual
Population = CostFunction(OPTIONS, Population);
% Sort the population from most fit to least fit
Population = PopSort(Population);
% Compute the average cost
AverageCost = ComputeAveCost(Population);
% Display info to screen
MinCost = [Population(1).cost];
AvgCost = [AverageCost];
if DisplayFlag
    disp(['The best and mean of Generation # 0 are ', num2str(MinCost(end)), ' and ',
num2str(AvgCost(end))]);
end

return;
```

ClearDups - This is used by each optimization method to get rid of duplicate population members and replace them with randomly generated individuals.

```
function [Population] = ClearDups(Population, MaxParValue, MinParValue)

% Make sure there are no duplicate individuals in the population.
% This logic does not make 100% sure that no duplicates exist, but any duplicates
that are found are
% randomly mutated, so there should be a good chance that there are no duplicates
after this procedure.
for i = 1 : length(Population)
    Chrom1 = sort(Population(i).chrom);
    for j = i+1 : length(Population)
        Chrom2 = sort(Population(j).chrom);
        if isequal(Chrom1, Chrom2)
            parnum = ceil(length(Population(j).chrom) * rand);
            Population(j).chrom(parnum) = floor(MinParValue + (MaxParValue -
MinParValue + 1) * rand);
        end
    end
end
return;
```

ComputeAveCost - This is used by each optimization method to compute the average cost of the population and to count the number of legal (feasible) individuals.

```
function [AveCost, nLegal] = ComputeAveCost(Population)
```

% Compute the average cost of all legal individuals in the population.
% OUTPUTS: AveCost = average cost
%   nLegal = number of legal individuals in population

```
% Save valid population member fitnesses in temporary array
Cost = [];
nLegal = 0;
for i = 1 : length(Population)
    if Population(i).cost < inf
         Cost = [Cost Population(i).cost];
         nLegal = nLegal + 1;
      end
end
% Compute average cost.
AveCost = mean(Cost);
return;
```

PopSort - This is used by each optimization method to sort population members from most fit to least fit.

```
function [Population, indices] = PopSort(Population)
```

```
% Sort the population members from best to worst
popsize = length(Population);
Cost = zeros(1, popsize);
indices = zeros(1, popsize);
for i = 1 : popsize
   Cost(i) = Population(i).cost;
end
[Cost, indices] = sort(Cost, 2, 'ascend');
Chroms = zeros(popsize, length(Population(1).chrom));
for i = 1 : popsize
   Chroms(i, :) = Population(indices(i)).chrom;
end
for i = 1 : popsize
   Population(i).chrom = Chroms(i, :);
   Population(i).cost = Cost(i);
end
```

Conclude - This is concludes the processing of each optimization method. It does common processing like outputting results.

```
function Conclude(DisplayFlag, OPTIONS, Population, nLegal, MinCost)
```

% Output results of population-based optimization algorithm.

```
if DisplayFlag
  % Count the number of duplicates
  NumDups = 0;
  for i = 1 : OPTIONS.popsize
    Chrom1 = sort(Population(i).chrom);
    for j = i+1 : OPTIONS.popsize
      Chrom2 = sort(Population(j).chrom);
      if isequal(Chrom1, Chrom2)
        NumDups = NumDups + 1;
      end
    end
  end
  disp([num2str(NumDups), ' duplicates in final population.']);
  disp([num2str(nLegal), ' legal individuals in final population.']);
  % Display the best solution
  Chrom = sort(Population(1).chrom);
  disp(['Best chromosome = ', num2str(Chrom)]);
  % Plot some results
  close all;
  plot([0:OPTIONS.Maxgen], MinCost, 'r');
  xlabel('Generation');
  ylabel('Minimum Cost');
end
return;
```

MAPSS - This is the sensor selection initialization and fitness evaluation function. It requires the Control System Toolbox. You can use any of the optimization algorithms to find an optimal sensor set by typing, for example, the following at the Matlab prompt:

```
>> PSO(@MAPSS);
function [InitFunction, CostFunction, FeasibleFunction] = MAPSS

% The following was found by exhaustive search to be the best 20/4 MAPSS
sensor set.
% However, this is computer-dependent because of numerical issues in Matlab's
DARE routine.
% [1 2 2 2 2 3 3 6 7 7 7 7 8 9 9 9 9 10 10 10]

InitFunction = @MAPSSInit;
CostFunction = @MAPSSCost;
FeasibleFunction = @MAPSSFeasible;
return;

%%%%%%%%%%%%%%%%%%%%%%%%%%%%%%%%%%%%%%%%%%%%
%%%%%%%%%%%%%%%%%%%%%%%%%%%%%%%%%%%
```

```
function [MaxParValue, MinParValue, Population, OPTIONS] =
MAPSSInit(OPTIONS)

global MaxParValue MinParValue NumDups A C Q R P0 alpha
NumDups = 4; % number of duplicates of each sensor that are allowed
alpha = 1; % relative importance of financial cost to estimation error
% Get MAPSS linearized system matrices A, C, Q, and R
load matrices.mat;
A = Aaug; C = Caug;
% Compute the reference steady state estimation error covariance
P0 = dare(A', C', Q, R, zeros(size(C')), eye(size(A)));
% Initialize population
for popindex = 1 : OPTIONS.popsize
    chrom = randperm(11 * NumDups);
    chrom = chrom(1 : OPTIONS.numVar);
    chrom = mod(chrom, 11);
    chrom(chrom==0) = 11;
    Population(popindex).chrom = chrom;
end
% Chromosome parameter can be any integer between 1 and 11 (sensor numbers)
MinParValue = 1;
MaxParValue = 11;
OPTIONS.OrderDependent = false;
return;

%%%%%%%%%%%%%%%%%%%%%%%%%%%%%%%%%%%%%%%%%%%
%%%%%%%%%%%%%%%%%%%%%%%%%%%%%%%%%%%%%%
function [Population] = MAPSSCost(OPTIONS, Population)

% Compute the sensor selection cost function of each member in Population
% Cost = sum(sqrt(P(i,i) / Pref(i,i))) + alpha * FinCost / RefFinCost
% There are 11 unique sensors that can be used in each sensor set.

global MaxParValue MinParValue NumDups A C Q R P0 alpha
popsize = OPTIONS.popsize;
DollarCost = 1000 * ones(11, 1); % dollar cost for initial use of each sensor
AdditionalCost = 750 * ones(11, 1); % dollar cost for duplicate sensors beyond the
first of each type
ReferenceCost = sum(DollarCost); % dollar cost if 11 unique sensors are used
for popindex = 1 : popsize
    New_Sensor_Set = Population(popindex).chrom;
    New_Sensor_Set = mod(New_Sensor_Set, 11);
    New_Sensor_Set(New_Sensor_Set==0) = 11;
    New_Sensor_Set = sort(New_Sensor_Set);
    %MANIPULATING C AND R MATRICES BASED ON RANDOMLY
GENERATED SENSOR
    %COMBINATION AND MAKING THE REMAINING ROWS ZEROS
    FIN_COST = 0;
```

```
for i = 1 : 11
    SENSOR(i).COUNT = 0;
end
for i = 1 : 11
  SENSOR(i).COUNT = length(find(New_Sensor_Set == i));
  if  SENSOR(i).COUNT > 0
      FIN_COST = FIN_COST + DollarCost(i); % initial sensor cost is defined
in DollarCost array
      FIN_COST = FIN_COST + (SENSOR(i).COUNT - 1) *
AdditionalCost(i);
  end
end
C_NEW = [];
R_NEW = [];
for i = 1 : length(New_Sensor_Set)
    SENSOR_NUM = New_Sensor_Set(i);
    C_NEW(i, :) = C(SENSOR_NUM, :);
    R_NEW(i, :) = R(SENSOR_NUM, SENSOR_NUM);
end
R_NEW = diag(R_NEW);
% Compute the steady state estimation error covariance based on the sensors
that are used
lastwarn('');
warning('off', 'control:InaccurateSolution');
[P_ss, L, G, REPORT] = dare(A', C_NEW', Q, R_NEW, zeros(size(C_NEW')),
eye(size(A)), 'report');
% If a steady state ARE solution does not exist, set the cost to a large number
if ~isempty(lastwarn) | REPORT == -1 | REPORT == -2
    Population(popindex).cost = 10e10;
    continue;
end
% Compute the cost of the sensor set: estimation error variance plus financial
cost
New_cost = 0;
for i = 4 : 11 % health parameters are indices 4-11 in the augmented state vector
    New_cost = New_cost + sqrt(P_ss(i,i) / P0(i,i));
end
New_cost = New_cost + alpha * FIN_COST / ReferenceCost;
Population(popindex).cost = New_cost;
if (New_cost <= 0) | ~isreal(New_cost) | (New_cost >= 100)
  New_cost = inf;
  end
  Population(popindex).cost = New_cost;
end
return
%%%%%%%%%%%%%%%%%%%%%%%%%%%%%%%%%%%%%%%%%%%
%%%%%%%%%%%%%%%%%%%%%%%%%%%%%%
```

```
function [Population] = MAPSSFeasible(OPTIONS, Population)

% Make sure each sensor set does not contain more than the allowable number of
copies of each sensor.

global MaxParValue MinParValue NumDups A C Q R P0 alpha
% Make sure none of the chromosomes has more than the allowable number of
sensors
i = 0;
while i < OPTIONS.popsize
    i = i + 1;
    Chrom = Population(i).chrom;
    for j = 1 : 11
        indices = find(Chrom == j);
        if length(indices) > NumDups
            % The individual has too many copies of a single sensor, so
            % replace the individual with a random sensor set.
            Chrom = randperm(11 * NumDups);
            Chrom = Chrom(1 : OPTIONS.numVar);
            Population(i).chrom = Chrom;
            i = i - 1; % decrement i so that this new individual can be checked again
            break;
        end
    end
end
% Make sure each chromosome is an integer between the allowable values
for i = 1 : OPTIONS.popsize
    Chrom = round(Population(i).chrom);
    Chrom = mod(Chrom, 11);
    Chrom(Chrom==0) = 11;
    Chrom = max(Chrom, MinParValue);
    Chrom = min(Chrom, MaxParValue);
    Population(i).chrom = Chrom;
end
return;
```

matrices - This is used by MAPSS.m and contains linearized system matrices for fitness function evaluation.

Monte - This can be used to obtain Monte Carlo simulation results. The first executable line specifies the number of simulations to run. This is the highest-level program in this archive, and is the one that I ran to create the results in the paper that I wrote.

```
function [MeanMin, MeanMinNorm, BestMin, BestMinNorm, MeanCPU] =
Monte

% Monte Carlo execution of population-based optimization software
% OUTPUT MeanMin is the mean of the best solution found. It is a
% nFunction x nBench array, where nFunction is the number of optimization
% functions that are used, and nBench is the number of benchmarks that
% are optimized.
% OUTPUT MeanMinNorm is MeanMin normalized to a minimum of 1 for each
benchmark.
% OUTPUT BestMin is the best solution found by each optimization function
% for each benchmark.
% OUTPUT BestMinNorm is BestMin normalized to a minimum of 1 for each
benchmark.
% OUTPUT MeanCPU is the mean CPU time required for each optimization
function
% normalized to 1.

nMonte = 100; % number of Monte Carlo runs

% Optimization methods
OptFunction = [
'ACO    '; % ant colony optimization
'BBO    '; % biogeography-based optimization
'DE     '; % differential evolution
'ES     '; % evolutionary strategy
'GA     '; % genetic algorithm
'PBIL   '; % probability based incremental learning
'PSO    '; % particle swarm optimization
'StudGA']; % stud genetic algorithm

% Benchmark functions
Bench = [      %        multimodal? separable?  regular?
'Ackley    '; %    y              n              y
'Fletcher  '; %    y              n              n
'Griewank  '; %    y              n              y
'Penalty1  '; %    y              n              y
'Penalty2  '; %    y              n              y
'Quartic   '; %    n              y              y
'Rastrigin '; %    y              y              y
'Rosenbrock'; %    n              n              y
'Schwefel  '; %    y              y              n
'Schwefel2 '; %    n              n              y
'Schwefel3 '; %    y              n              n
'Schwefel4 '; %    n              n              n
'Sphere    '; %    n              y              y
'Step      ']; %   n              y              n
```

```
%Bench = ['MAPSS'];

nFunction = size(OptFunction, 1);
nBench = size(Bench, 1);
MeanMin = zeros(nFunction, nBench);
BestMin = inf(nFunction, nBench);
MeanCPU = zeros(nFunction, nBench);
for i = 1 : nFunction
    for j = 1 : nBench
        disp(['Optimization method ', num2str(i), '/', num2str(nFunction), ...
            ', Benchmark function ', num2str(j), '/', num2str(nBench)]);
        for k = 1 : nMonte
            tic;
            [Cost] = eval([OptFunction(i,:), '(@', Bench(j,:), ', false);']);
            MeanCPU(i,j) = ((k - 1) * MeanCPU(i,j) + toc) / k;
            MeanMin(i,j) = ((k - 1) * MeanMin(i,j) + Cost(end)) / k;
            BestMin(i,j) = min(BestMin(i,j), Cost(end));
        end
    end
end
% Normalize the results
if min(MeanMin) == 0
    MeanMinNorm = [];
else
    MeanMinNorm = MeanMin * diag(1./min(MeanMin));
end
if min(BestMin) == 0
    BestMinNorm = [];
else
    BestMinNorm = BestMin * diag(1./min(BestMin));
end
MeanCPU = min(MeanCPU');
MeanCPU = MeanCPU / min(MeanCPU);
```

Finaly the BBO Algorithm

```
function [MinCost, Hamming] = BBO(ProblemFunction, DisplayFlag, ProbFlag,
RandSeed)

% Biogeography-based optimization (BBO) software for minimizing a general
function

% INPUTS: ProblemFunction is the handle of the function that returns
%         the handles of the initialization, cost, and feasibility functions.
```

```
%         DisplayFlag = true or false, whether or not to display and plot results.
%         ProbFlag = true or false, whether or not to use probabilities to update
emigration rates.
%         RandSeed = random number seed
% OUTPUTS: MinCost = array of best solution, one element for each generation
%         Hamming = final Hamming distance between solutions
% CAVEAT: The "ClearDups" function that is called below replaces duplicates
with randomly-generated
%         individuals, but it does not then recalculate the cost of the replaced
individuals.

if ~exist('DisplayFlag', 'var')
    DisplayFlag = true;
end
if ~exist('ProbFlag', 'var')
    ProbFlag = false;
end
if ~exist('RandSeed', 'var')
    RandSeed = round(sum(100*clock));
end

[OPTIONS, MinCost, AvgCost, InitFunction, CostFunction, FeasibleFunction, ...
    MaxParValue, MinParValue, Population] = Init(DisplayFlag, ProblemFunction,
RandSeed);

Population = CostFunction(OPTIONS, Population);

OPTIONS.pmodify = 1; % habitat modification probability
OPTIONS.pmutate = 0.005; % initial mutation probability

Keep = 2; % elitism parameter: how many of the best habitats to keep from one
generation to the next
lambdaLower = 0.0; % lower bound for immigration probabilty per gene
lambdaUpper = 1; % upper bound for immigration probabilty per gene
dt = 1; % step size used for numerical integration of probabilities
I = 1; % max immigration rate for each island
E = 1; % max emigration rate, for each island
P = OPTIONS.popsize; % max species count, for each island

% Initialize the species count probability of each habitat
% Later we might want to initialize probabilities based on cost
for j = 1 : length(Population)
    Prob(j) = 1 / length(Population);
end
```

```
% Begin the optimization loop
for GenIndex = 1 : OPTIONS.Maxgen
    % Save the best habitats in a temporary array.
    for j = 1 : Keep
        chromKeep(j,:) = Population(j).chrom;
        costKeep(j) = Population(j).cost;
    end
    % Map cost values to species counts.
    [Population] = GetSpeciesCounts(Population, P);
    % Compute immigration rate and emigration rate for each species count.
    % lambda(i) is the immigration rate for habitat i.
    % mu(i) is the emigration rate for habitat i.
    [lambda, mu] = GetLambdaMu(Population, I, E, P);
    if ProbFlag
        % Compute the time derivative of Prob(i) for each habitat i.
        for j = 1 : length(Population)
            % Compute lambda for one less than the species count of habitat i.
            lambdaMinus = I * (1 - (Population(j).SpeciesCount - 1) / P);
            % Compute mu for one more than the species count of habitat i.
            muPlus = E * (Population(j).SpeciesCount + 1) / P;
            % Compute Prob for one less than and one more than the species count of
habitat i.
            % Note that species counts are arranged in an order opposite to that
presented in
            % MacArthur and Wilson's book - that is, the most fit
            % habitat has index 1, which has the highest species count.
            if j < length(Population)
                ProbMinus = Prob(j+1);
            else
                ProbMinus = 0;
            end
            if j > 1
                ProbPlus = Prob(j-1);
            else
                ProbPlus = 0;
            end
            ProbDot(j) = -(lambda(j) + mu(j)) * Prob(j) + lambdaMinus * ProbMinus +
muPlus * ProbPlus;
        end
        % Compute the new probabilities for each species count.
        Prob = Prob + ProbDot * dt;
        Prob = max(Prob, 0);
        Prob = Prob / sum(Prob);
    end
    % Now use lambda and mu to decide how much information to share between
habitats.
```

```
lambdaMin = min(lambda);
lambdaMax = max(lambda);
for k = 1 : length(Population)
   if rand > OPTIONS.pmodify
     continue;
   end
     % Normalize the immigration rate.
     lambdaScale = lambdaLower + (lambdaUpper - lambdaLower) * (lambda(k)
- lambdaMin) / (lambdaMax - lambdaMin);
   % Probabilistically input new information into habitat i
   for j = 1 : OPTIONS.numVar
     if rand < lambdaScale
        % Pick a habitat from which to obtain a feature
        RandomNum = rand * sum(mu);
        Select = mu(1);
        SelectIndex = 1;
        while (RandomNum > Select) & (SelectIndex < OPTIONS.popsize)
           SelectIndex = SelectIndex + 1;
           Select = Select + mu(SelectIndex);
         end
        Island(k,j) = Population(SelectIndex).chrom(j);
        else
           Island(k,j) = Population(k).chrom(j);
     end
   end
end
if ProbFlag
   % Mutation
   Pmax = max(Prob);
   MutationRate = OPTIONS.pmutate * (1 - Prob / Pmax);
   % Mutate only the worst half of the solutions
   Population = PopSort(Population);
   for k = round(length(Population)/2) : length(Population)
      for parnum = 1 : OPTIONS.numVar
         if MutationRate(k) > rand
            Island(k,parnum) = floor(MinParValue + (MaxParValue -
MinParValue + 1) * rand);
      end
    end
   end
end
% Replace the habitats with their new versions.
for k = 1 : length(Population)
   Population(k).chrom = Island(k,:);
end
% Make sure each individual is legal.
Population = FeasibleFunction(OPTIONS, Population);
% Calculate cost
```

```
Population = CostFunction(OPTIONS, Population);
% Sort from best to worst
Population = PopSort(Population);
% Replace the worst with the previous generation's elites.
n = length(Population);
for k = 1 : Keep
    Population(n-k+1).chrom = chromKeep(k,:);
    Population(n-k+1).cost = costKeep(k);
end
  % Make sure the population does not have duplicates.
  Population = ClearDups(Population, MaxParValue, MinParValue);
  % Sort from best to worst
  Population = PopSort(Population);
  % Compute the average cost
  [AverageCost, nLegal] = ComputeAveCost(Population);
  % Display info to screen
  MinCost = [MinCost Population(1).cost];
  AvgCost = [AvgCost AverageCost];
  if DisplayFlag
    disp(['The best and mean of Generation # ', num2str(GenIndex), ' are ',...
      num2str(MinCost(end)), ' and ', num2str(AvgCost(end))]);
  end
end
Conclude(DisplayFlag, OPTIONS, Population, nLegal, MinCost);
% Obtain a measure of population diversity
for k = 1 : length(Population)
  Chrom = Population(k).chrom;
  for j = MinParValue : MaxParValue
    indices = find(Chrom == j);
    CountArr(k,j) = length(indices); % array containing gene counts of each
habitat
  end
end
Hamming = 0;
for m = 1 : length(Population)
  for j = m+1 : length(Population)
    for k = MinParValue : MaxParValue
      Hamming = Hamming + abs(CountArr(m,k) - CountArr(j,k));
    end
  end
end
if DisplayFlag
  disp(['Diversity measure = ', num2str(Hamming)]);
end
return;
%%%%%%%%%%%%%%%%%%%%%%%%%%%%%%%%%%%%%%%%%
%%%%%%%%%%%%%%%%%%%%%%%%%%%%%
```

```
function [Population] = GetSpeciesCounts(Population, P)

% Map cost values to species counts.

% This loop assumes the population is already sorted from most fit to least fit.
for i = 1 : length(Population)
    if Population(i).cost < inf
      Population(i).SpeciesCount = P - i;
    else
      Population(i).SpeciesCount = 0;
    end
end
return;

%%%%%%%%%%%%%%%%%%%%%%%%%%%%%%%%%%%%%%%%%%%
%%%%%%%%%%%%%%%%%%%%%%%%%%%%%%%%
function [lambda, mu] = GetLambdaMu(Population, I, E, P)

% Compute immigration rate and extinction rate for each species count.
% lambda(i) is the immigration rate for individual i.
% mu(i) is the extinction rate for individual i.

for i = 1 : length(Population)
    lambda(i) = I * (1 - Population(i).SpeciesCount / P);
    mu(i) = E * Population(i).SpeciesCount / P;
end
return;
```

Appendix D: *Sample Data Input and Output*

The sample files are given in this section.

1. The Compatibility matrix of 21 cells is given as follows. Here the cell value is '5' when compatibility is checked amongst the channels in the same cell, the value is '3' when compared among the channels in 2nd nearest neighbor and '2' when compared in next neighboring cell. It is '0' when the cell is beyond 2nd next.

```
5 3 2 0 0 2 3 3 2 0 0 0 0 2 2 2 0 0 0 0 0
3 5 3 2 0 0 2 3 3 2 0 0 0 0 2 2 2 0 0 0 0
2 3 5 3 2 0 0 2 3 3 3 0 0 0 0 2 2 2 0 0 0
0 2 3 5 3 0 0 0 2 3 3 2 0 0 0 0 2 2 0 0 0
0 0 2 3 5 0 0 0 0 2 3 3 0 0 0 0 0 2 0 0 0
2 0 0 0 0 5 3 2 0 0 0 0 3 3 2 0 0 0 0 0 0
3 2 0 0 0 3 5 3 2 0 0 0 2 3 3 2 0 0 2 0 0
3 3 2 0 0 2 3 5 3 2 0 0 0 2 0 0 2 0 2 2 0
2 3 3 2 0 0 2 3 5 3 2 0 0 0 2 3 3 2 2 2 2
0 2 3 3 2 0 0 2 0 5 3 2 0 0 0 2 3 3 0 2 2
0 0 2 3 2 0 0 0 2 3 5 3 0 0 0 0 2 3 0 0 2
0 0 0 2 3 0 0 0 0 2 3 5 0 0 0 0 0 2 0 0 0
0 0 0 0 0 3 2 0 0 0 0 0 5 3 2 0 0 0 0 0 0
2 0 0 0 0 3 3 2 0 0 0 0 3 5 3 2 0 0 2 0 0
2 2 0 0 0 2 3 3 2 0 0 0 2 3 5 3 2 0 3 2 0
2 2 2 0 0 0 2 3 3 2 0 0 0 2 3 5 3 2 3 3 2
0 2 2 2 0 0 0 2 3 3 2 0 0 2 3 5 3 2 3 3
0 0 2 2 2 0 0 0 2 3 3 2 0 0 2 3 5 0 2 3
0 0 0 0 0 0 2 2 2 0 0 0 0 2 3 3 2 0 5 3 2
0 0 0 0 0 0 0 2 2 2 0 0 0 0 2 3 3 2 3 5 3
0 0 0 0 0 0 0 0 2 2 2 0 0 0 0 2 3 3 2 3 5
```

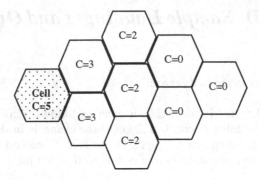

Fig. A1 Compatibility values in different cells

The following tables represent a sample of the hotness matrix and ots changes.

28 23 24 35 40 42 39 5 8 7 15 29 27 22 39 45 32 23 14 11 7

This is the hotness of 21 cells in a given time t= t_0
The subsequent changes of with respect to time is given as

66 43 65 97 82 77 75 89 130 145 0 146 44 95 25 49 103
133 135 5 46

145   60 69 148 30 39 13 33 98 95 80 136 30 38 116 149 112 139 57 95 72

58 29 36 148 1 31 148 25 51 5 15 86 143 124 93 51 74 134 117 24 78

3 132 81 3 101 105 130 9 5 40 55 96 22 37 83 11 20 76 45 0 92

36 12 62 61 43 79 101 103 134 119 11 82 89 149 114 33 75 14 96 132 98

81 21 52 104 18 50 70 84 21 55 99 58 88 107 46 69 44 21 72 37 111

143 108 61 84 62 46 132 62 122 122 123 26 28 48 91 36 25 13 88 122 126

13 128 146 31 27 111 71 19 50 29 140 36 105 129 76 125 131 148 89 20 34

146 111 146 95 99 39 52 62 30 109 106 127 93 136 98 28 137 25 120 49 113

18 63 25 95 57 115 133 94 30 83 11 69 146 91 134 31 45 8 57 24 87

27 139 46 123 31 6 136 114 30 86 89 122 56 48 87 2 130 52 37 83 67

121 118 136 13 109 34 41 134 71 60 50 55 50 30 143 66 146 120 138 45 32

4 45 60 138 104 78 105 41 144 117 87 65 93 126 50 118 58 129 127 106 149

37 42 16 37 46 49 112 125 89 145 65 7 99 30 34 20 43 112 109 82 115

19 73 117 56 26 116 148 79 50 91 25 128 128 50 5 101 38 99 63 2 146

65 43 113 132 8 31 149 56 121 63 31 102 5 35 55 49 62 149 78 22 38

48 144 137 16 124 91 149 81 15 103 0 68 5 34 143 110 58 89 129 94 100

82 34 99 19 147 66 67 5 132 78 37 149 101 30 86 80 71 113 114 88 61

96 115 36 73 51 21 44 48 61 51 5 5 24 75 111 95 6 101 56 61 10

108 134 55 52 135 106 60 18 29 113 36 57 101 76 112 55 99 107 106 15 21

135 103 42 30 79 8 88 136 143 105 123 51 132 105 135 42 88 45 17 42 92

83 94 76 16 147 115 36 87 80 143 34 48 47 121 144 12 34 63 149 54 20

32 45 144 92 49 143 7 130 20 144 53 145 103 95 142 144 101 49 148 134 129

The following table represents the neighbour information of the cells. The rows represent 21 cells and column value gives its six neighbors. The 1st row gives describe the 1st cell that has no cell in 1st 2nd side. Its 3rd side has 7th cell 4th has 8th cell and 5th has 2nd cell. Again 6th side has no neighbour.

0	0	7	8	2	0
0	1	8	9	3	0
0	2	9	10	4	0
0	3	10	11	5	0
0	4	11	12	0	0
0	0	13	14	7	0
0	6	14	15	8	1
1	7	15	16	9	2
2	8	16	17	10	3
3	9	17	18	11	4
4	10	18	0	12	5
5	11	0	0	0	0
0	0	0	0	14	6
6	13	0	0	15	7
7	14	0	19	16	8
8	15	19	20	17	9
9	16	20	21	18	10
10	17	21	0	0	11
15	0	0	0	20	16
16	19	0	0	21	17
17	20	0	0	0	18

Next we take the data for the velocity of the mobile stations in each cells. We assume all the MS are moving and generate randomly a velocity value in the interval(0,140). This is chosen keeping in mind the city traffic.

23	80	77	55	9	9	25	23	97	33	1	1	54	82	103	27	104	52	14	35	104
24	100	104	74	110	110	66	92	92	90	10	70	55	4	98	83	21	6	54	37	22
66	92	13	96	56	79	33	51	30	24	30	104	41	14	91	27	73	15	65	79	66
60	25	79	39	27	112	23	38	12	95	12	53	43	75	22	29	56	98	114	15	41
88	61	113	40	1	79	11	69	7	108	25	63	67	58	90	61	73	2	17	23	22
95	41	15	81	73	27	69	64	92	37	39	48	98	69	58	66	102	97	51	98	82
61	9	17	57	41	79	47	104	29	16	25	36	43	57	50	24	73	24	67	55	103

Here the column represents all 21 cells and rows are MS assigned in each channel in the cell.

The 1st 15 parrents are taken as

parent 0 :	parent 1 :	parent 2 :	parent 3 :	parent 4 :
0001010	0110100	0000110	1110101	0101001
1011110	0100111	1011011	0011001	1110010
0110010	1110001	1010100	0001110	0101100
1000000	0110111	1011011	0101101	1110101
1011101	0001000	1001101	0000011	0111001
0000011	1100100	0010001	1000101	0110001
1101111	0011111	0010001	0011101	0000100
0010011	0001100	0100110	0000111	1111110
1000110	1001011	0011010	1110110	0101000
1111000	0101111	0110110	0101010	1100111
0110001	0101010	0011011	1111000	1001110
1000000	1010110	1011110	0000111	1011100
0010100	1011000	0000111	1111001	0100111
1100111	0010000	0001101	1111100	0000010
1010110	1101000	1000101	1110110	0101001
1000110	0100100	1001011	0011010	1101011
1111011	0101101	0001000	1000100	0000011
0101100	0101100	1000110	0010100	0001001
1111110	1110101	1111001	0110011	1111110
0100011	1101000	0110000	0011001	0101101
0101001	0010011	0010011	0111100	0100001

parent 5 :	parent 6 :	parent 7 :	parent 8 :	parent 9 :
0100010	0101111	0000101	1111011	0101100
0010001	1001110	0001011	0100101	1001011
1101111	0010001	1000011	0000111	0100000
0110010	1100011	1000101	1100111	1110000
1001101	0110010	1000001	1111111	1011111
1001001	1110001	0100001	1000111	0101100
1111101	1011101	1110000	1011000	1111111
1111000	0111000	0001011	1001010	1001110
1100100	0000010	0110111	0010011	0111000
1110010	1001110	0010000	1000111	0110010
1011111	0010111	0001011	1100110	1001101
0011110	1110011	0111011	1001100	0111110
0001111	0101111	1010010	1111000	0110001
0100101	1001001	0111011	0110010	0001100
0101101	0000100	1010101	0000000	0001011
1011011	1001010	0011100	0000000	0111011
0001001	0111111	0110010	1011101	1101011
1001100	1101110	0100011	0001101	1101111
1000101	0001010	0100001	0111000	0011100
1001000	0000111	1100110	0010011	1000101
1011110	0010000	1011111	1000010	0101001

parent 10 :	parent 11 :	parent 12 :	parent 13 :	parent 14 :
0010001	1110011	1100001	1100010	1010101
0010100	1101010	0011111	0110101	1101011
0001010	0011110	0010011	0001001	1111001
0010100	0110100	0001110	0000100	0101010
1101011	0111111	1111001	0000000	1101111
1101011	0000110	0101000	1100101	0101011
0010000	1111001	0010011	1110000	0011100
1100100	1111100	0100000	0111001	0100001
0001110	1000011	1000011	0101110	1111110
1010111	1101100	0010111	0001110	0001101
0111111	1101011	1010011	0100000	0010111
0101001	0100001	1010101	0000001	1100101
0000111	1101101	1011011	0001100	1111000
0100101	1010011	0011110	0000111	1100010
1011011	1001011	0001111	0110010	1101001
0001100	1001011	0000110	0001011	1101111
1011010	1000110	1011100	0010111	1010110
0010100	0010011	1110001	0000100	0000000
1011111	1001001	0111011	0111111	1011010
0111110	1010000	0000001	1000011	1011011
0010011	0110100	1000001	0111101	0110100

The angles in which the MS are moving w.r.t. BS are generated randomly. They are given as

Cell 1	Cell 5	Cell 10	Cell 15	Cell 20
22.600	76.394	73.529	52.521	9.2309
23.554	95.493	99.949	70.983	105.67
63.343	88.490	12.732	91.673	53.794
57.932	23.873	75.439	37.242	26.101
84.033	58.887	108.54	38.833	1.5915
90.718	39.788	14.642	77.667	70.346
58.250	8.5943	16.552	54.749	39.4704

The Offspring generated Using *Genetic Algorithm* technique are given below at any time $t=t_0$

t_0 th offspring	t_1 th offspring	t_2 th offspring	t_3 th offspring
0 0 0 1 0 1 0	0 1 1 0 1 0 0	0 0 0 0 1 1 0	1 1 1 0 1 0 1
1 0 1 1 1 1 0	0 1 0 0 1 1 1	1 0 1 1 0 1 1	0 0 1 1 0 0 1
0 1 1 0 0 1 0	1 1 1 0 0 0 1	1 0 1 0 1 0 0	0 0 0 1 1 1 0
1 0 0 0 0 0 0	0 1 1 0 1 1 1	1 0 1 1 0 1 1	0 1 0 1 1 0 1
1 0 1 1 0 0 1	0 0 0 1 0 0 0	1 0 0 1 1 0 1	0 0 0 0 0 1 1
0 0 0 0 0 1 0	1 1 0 0 1 0 0	0 0 1 0 0 0 1	1 0 0 0 1 0 1
1 1 0 1 0 1 0	0 0 1 1 1 1 1	0 0 1 0 0 0 1	0 0 1 1 1 0 1
0 0 1 0 0 1 1	0 0 0 1 1 0 0	0 1 0 0 1 1 0	0 0 0 0 1 1 1
1 0 0 0 1 1 0	1 0 0 1 0 1 1	0 0 1 1 0 1 0	1 1 1 0 1 1 0
1 1 0 1 0 0 0	0 1 0 1 1 1 1	0 1 1 0 1 1 0	0 1 0 1 0 1 0
0 1 0 0 0 0 1	0 1 0 1 0 1 0	0 0 1 1 0 1 1	1 1 1 1 0 0 0
1 0 0 0 0 0 0	1 0 1 0 1 1 0	1 0 1 1 1 1 0	0 0 0 0 1 1 1
0 0 1 0 1 0 0	1 0 1 1 0 0 0	0 0 0 0 1 1 1	1 1 1 1 0 0 1
1 1 0 0 1 1 1	0 0 1 0 0 0 0	0 0 0 1 1 0 1	1 1 1 1 1 0 0
1 0 1 0 1 1 0	1 1 0 1 0 0 0	1 0 0 0 1 0 1	1 1 1 0 1 1 0
1 0 0 0 1 1 0	0 1 0 0 1 0 0	1 0 0 1 0 1 1	0 0 1 1 0 1 0
1 0 1 1 0 1 1	0 1 0 1 1 0 1	0 0 0 1 0 0 0	1 0 0 0 1 0 0
0 1 0 1 1 0 0	0 1 0 1 1 0 0	1 0 0 0 1 1 0	0 0 1 0 1 0 0
1 1 1 0 1 1 0	1 1 1 0 1 0 1	1 1 1 1 0 0 1	0 1 1 0 0 1 1
0 0 0 0 0 1 1	1 1 0 1 0 0 0	0 1 1 0 0 0 0	0 0 1 1 0 0 1
0 0 0 1 0 0 1	0 0 1 0 0 1 1	0 0 1 0 0 1 1	0 1 1 1 1 0 0

The Offspring generated Using *Biogeography Based Optimization* technique are given below at any time $t=t_0$

0 th offspring	1 th offspring	2 th offspring	3 th offspring
0 0 0 1 0 1 0	0 1 1 0 1 0 0	0 0 0 0 1 1 0	1 1 1 0 1 0 1
1 0 1 1 1 1 0	0 1 0 0 1 1 1	1 0 1 1 0 1 1	0 0 1 1 0 0 1
0 1 1 0 0 1 0	1 1 1 0 0 0 1	1 0 1 0 1 0 0	0 0 0 1 1 1 0
1 0 0 0 0 0 0	0 1 1 0 1 1 1	1 0 1 1 0 1 1	0 1 0 1 1 0 1
1 0 1 1 0 0 1	0 0 0 1 0 0 0	1 0 0 1 1 0 1	0 0 0 0 0 1 1
0 0 0 0 0 1 0	1 1 0 0 1 0 0	0 0 1 0 0 0 1	1 0 0 0 1 0 1
1 1 0 1 0 1 0	0 0 1 1 1 1 1	0 0 1 0 0 0 1	0 0 1 1 1 0 1
0 0 1 0 0 1 1	0 0 0 1 1 0 0	0 1 0 0 1 1 0	0 0 0 0 1 1 1
1 0 0 0 1 1 0	1 0 0 1 0 1 1	0 0 1 1 0 1 0	1 1 1 0 1 1 0
1 1 0 1 0 0 0	0 1 0 1 1 1 1	0 1 1 0 1 1 0	0 1 0 1 0 1 0
0 1 0 0 0 0 1	0 1 0 1 0 1 0	0 0 1 1 0 1 1	1 1 1 1 0 0 0
1 0 0 0 0 0 0	1 0 1 0 1 1 0	1 0 1 1 1 1 0	0 0 0 0 1 1 1
0 0 1 0 1 0 0	1 0 1 1 0 0 0	0 0 0 0 1 1 1	1 1 1 1 0 0 1
1 1 0 0 1 1 1	0 0 1 0 0 0 0	0 0 0 1 1 0 1	1 1 1 1 1 0 0
1 0 1 0 1 1 0	1 1 0 1 0 0 0	1 0 0 0 1 0 1	1 1 1 0 1 1 0
1 0 0 0 1 1 0	0 1 0 0 1 0 0	1 0 0 1 0 1 1	0 0 1 1 0 1 0
1 0 1 1 0 1 1	0 1 0 1 1 0 1	0 0 0 1 0 0 0	1 0 0 0 1 0 0
0 1 0 1 1 0 0	0 1 0 1 1 0 0	1 0 0 0 1 1 0	0 0 1 0 1 0 0
1 1 1 0 1 1 0	1 1 1 0 1 0 1	1 1 1 1 0 0 1	0 1 1 0 0 1 1
0 0 0 0 0 1 1	1 1 0 1 0 0 0	0 1 1 0 0 0 0	0 0 1 1 0 0 1
0 0 0 1 0 0 1	0 0 1 0 0 1 1	0 0 1 0 0 1 1	0 1 1 1 1 0 0

Index

Authors

Sanchita Ghosh is currently an Assistant Professor in the Department of Computer Science and Engineering, Birla Institute of Technology, Mesra. She is also a Visiting Faculty of Jadavpur University, where she offers post graduate level course in Image Processing. Sanchita received her M. Sc degree in Mathematics from Indian Institute of Technology, Kharagpur and M. Tech degree in Computer Science from Birla Institute of Technology, Mesra. She has done her Ph.D. from Jadavpur University.

Amit Konar is currently a Professor in the dept. of Electronics and Tele-communication Engineering (ETCE), Jadavpur University. He is founding coordinator of the M.Tech. program on Intelligent Automation and Robotics, offered by ETCE department, Jadavpur University. He received his B. E degree from Bengal Engineering and Science University (B. E. College), Shibpur in 1983 and his M. E. Tel E, M. Phil. and Ph.D. (Engineering) degrees from Jadavpur University in 1985, 1988, and 1994 respectively. Dr. Konar's research areas include the study of computational intelligence algorithms and their applications 320 Author's Biography to the entire domain of electrical engineering and computer science. Specifically he worked on sets and logic, neuro-computing, evolutionary algorithms, Dempster- Shafer theory and Kalman filtering, and applied the principles of computational intelligence in image understanding, VLSI design, mobile robotics and pattern recognition. Dr. Konar has supervised 10 Ph.D. theses. He has around 200 publications in international journal and conferences. He is an author of 6 books, including two popular texts: *Artificial Intelligence and Soft Computing*, from CRC Press in 2000 and *Computational Intelligence: Principles, Techniques and Applications* from Springer in 2005. Dr. Konar serves as the Editor-in-Chief of the *International Journal of Artificial Intelligence and Soft Computing* from Inderscience, U.K., and he is also the member of the editorial board of 5 other international journals. He was a recipient of AICTE-accredited

1997-2000 Career Award for Young Teachers for his significant contribution in teaching and research. He was a Visiting Professor for the Summer Courses in University of Missouri, St. Louis, USA in 2006. Dr. Konar is a Principal Investigator or Co-Principal Investigator of four external projects funded by University Grants Commission (the UGC is one of the main federal funding agencies in India) and two projects funded by the All India Council of Technical Education, Government of India. The research areas of these projects include decision support system for criminal investigation; navigational planning for mobile robots, AI and image processing, neural net based dynamic channel allocation, human mood detection from facial expressions and DNAstring matching algorithms. Dr. Konar served as a member of Program Committee of several International Conferences and workshops, such as Intl. Conf. on Hybrid Intelligent Systems (HIS 2003), held in Adelaide, Australia and Int. Workshop on Distributed Computing (IWDC 2002), held in Calcutta.